Climate Risk Management, Policy and Governance

This book series is devoted to the growing body of studies that provide analytical insight for policy-making and implementation for bridging climate change adaptation, disaster management and development sectors. It is reflective on all aspects of the climate risk management process, including assessment, mapping, identification, communication, implementation, governance and evaluation of climate risks and management responses.

Topics may span across global, national, regional, sectoral and local scales. The series invites multi-disciplinary and transdisciplinary approaches, combining insights from natural science, engineering and social sciences; emphasizing existing gaps, particularly in the area of decision-making, governance and international relations.

The series furthermore offers both theoretical and practical contributions, with the aim to further academic study and thinking, as well as advancing policy making and implementation of climate risk management processes and tools.

More information about this series at http://www.springer.com/series/15515

Swenja Surminski · Thomas Tanner
Editors

Realising the 'Triple Dividend of Resilience'

A New Business Case for Disaster Risk Management

Springer

Editors
Swenja Surminski
Grantham Research Institute on Climate Change and the Environment
London School of Economics
London
UK

Thomas Tanner
Overseas Development Institute
London
UK

ISSN 2510-1390 ISSN 2510-1404 (electronic)
Climate Risk Management, Policy and Governance
ISBN 978-3-319-40693-0 ISBN 978-3-319-40694-7 (eBook)
DOI 10.1007/978-3-319-40694-7

Library of Congress Control Number: 2016953656

Printed on acid-free paper

This Springer imprint is published by Springer Nature
The registered company is Springer International Publishing AG
The registered company address is: Gewerbestrasse 11, 6330 Cham, Switzerland

Foreword

This past decade witnessed nearly twice as many natural disasters as the 1980s. This number will continue to rise as urbanization progresses, and the effect of climate change becomes more apparent. To protect lives and livelihoods, it is important for communities to take precautionary action, *before* nature strikes. Greater investment in risk prevention and disaster preparedness can significantly lower losses and yield positive benefits for the society.

To incentivise greater investment in resilience interventions, the Global Facility for Disaster Reduction and Recovery (GFDRR) has been working with the World Bank and the Overseas Development Institute to strengthen the business case for investment in disaster risk management (DRM) and climate adaptation. The series of publications in this book articulate the business case for investment in DRM beyond avoided losses from disasters, explaining the multitude of development benefits generated by DRM activities. The synthesis report organizes these benefits into three categories or dividends:

Preventing loss of life and property: The fundamental reason for investing in disaster resilience will always be to save lives and avoid losses in the event of a disaster. In 1999, the most powerful tropical cyclone ever recorded in the North Indian Ocean claimed nearly 10,000 lives in Odisha, India. This triggered enormous efforts to reduce the risks associated with a similar disaster, and when an equally powerful storm hit in 2013, timely evacuation along the coast meant that fatalities were less than 1 % of the prior event's toll.

Attracting investment by lowering the looming threat of losses from disasters: Unlock economic potential by reducing background risk, which can restrict long-term investments in income-generating assets, entrepreneurial enterprises, and other growth areas. For example, Mexico's CADENA programme shows that weather-indexed insurance not only helps to compensate farmers for drought losses but also enables poor farmers to overcome credit constraints and invest in tools and fertilisers, boosting their productivity.

Capturing additional development benefits that are often 'unnoticed': Resilience measures often have multiple uses beyond disaster risk reduction. For example, the World Bank's flood management programme in Colombo, which includes preserving its wetlands, is playing a valuable role in reducing the city's flood risk. The preservation of wetlands is in itself yielding benefits beyond risk reduction such as economic security to residents through fishing and rice cultivation, a park for tourism and recreational activities, and lower temperatures in areas around the wetlands that on average remain 10° cooler that areas in the city.

The new 2030 development agenda, including the Sendai Framework for Disaster Risk Reduction, the Sustainable Development Goals (SDGs), and the Paris Declaration—all echo the increased awareness for the need to build disaster and climate resilience. Making the case for investing in disaster resilience is therefore important and timely, especially to ensure that all future development is risk-informed and therefore, sustainable. Strategically housed within the World Bank, GFDRR draws on a vast network of development and economic actors to support countries, understand their risk to natural hazards, and make informed decisions to build disaster resilience. Much remains to be done to scale up efforts to integrate risk reduction measures into development programmes and achieve the ambitions laid out for 2030. Highlighting the many benefits generated from investing in disaster resilience will play a central role in the process.

Francis Ghesquiere
Head of the GFDRR Secretariat, World Bank

Acknowledgements

This book is an output of an initiative funded by the Global Facility for Disaster Reduction and Recovery (GFDRR) at the World Bank and led by the Overseas Development Institute (ODI).

Our special thanks to Francis Geshquire and Tom Mitchell for initating discussions and establishing this work, and to Rob Reid and Sumati Rajput at the GFDRR for overseeing the work so professionally. We gratefully acknowledge the wide range of people at the Overseas Development Institute, World Bank and elsewhere that have provided advice, review, and inputs into the research. Our thanks to Lindsey Jones, Anna Locke, Emma Lovell, Tom Mitchell, Katie Peters, Catherine Simonet, Emily Wilkinson and Dirk Willem te Velde (all ODI), Javier Baez, Mook Bangalore, Carter Brandon, Oscar Anil Ishizawa Escudero, Marc S. Forni, Leonardo Garrido, Francis Ghesquiere, Rashmin Gunasekera, Stephane Hallegatte, Juan Jose Miranda Montero, Niels Holm-Nielsen, Jun Rentschler (all World Bank), Stephany Griffith-Jones (Columbia University), Marie-Agnes Jouanjean (OECD), Reinhard Mechler, Junko Mochizuki, Stefan Hochrainer-Stigler (all IIASA), Tom McDermott (University College Cork), Adam Rose (University of Southern California), Francis Vorhies (Earthmind).

Thanks also to Ore Kolade for efficient project management, Roo Griffiths for copyediting, and Mads Freund Brunse for figure designs.

Swenja Surminski

November 2016 Thomas Tanner

Contents

Contributors

Mook Bangalore Climate Change Group, World Bank, Washington, DC, USA

Stephany Griffith-Jones Columbia University, New York, USA

Stéphane Hallegatte Climate Change Group, World Bank, Washington, DC, USA

Stefan Hochrainer-Stigler International Institute for Applied Systems Analysis, Laxenburg, Austria

Marie-Agnès Jouanjean Organization for Economic Cooperation and Development, Paris, France

Emma Lovell Overseas Development Institute, London, UK

Thomas K.J. McDermott School of Economics, University College Cork, Cork, Ireland; Grantham Research Institute on Climate Change and the Environment, LSE, London, UK

Reinhard Mechler International Institute for Applied Systems Analysis, Laxenburg, Austria

Junko Mochizuki International Institute for Applied Systems Analysis, Laxenburg, Austria

Sumati Rajput Global Facility for Disaster Reduction and Recovery (GFDRR) at the World Bank, Washington, DC, USA

Robert Reid Global Facility for Disaster Reduction and Recovery (GFDRR) at the World Bank, Washington, DC, USA

Jun Rentschler Global Facility for Disaster Reduction and Recovery (GFDRR) at the World Bank, Washington, DC, USA

Adam Rose Price School of Public Policy and Center for Risk and Economic Analysis of Terrorism Events, University of Southern California, Los Angeles, CA, USA

Swenja Surminski Grantham Research Institute on Climate Change and the Environment, London School of Economics, London, UK

Thomas Tanner Overseas Development Institute, London, UK

Francis Vorhies Earthmind, Divonne-les-Bains, France

Emily Wilkinson Overseas Development Institute, London, UK

Abbreviations

ACP	African, Caribbean, and Pacific
ADB	Asian Development Bank
ASEAN	Association of Southeast Asian Nations
BI	Business Interruption
CAT	Catastrophe
CATSIM	Catastrophe Simulation
CBA	Cost–Benefit Analysis
CCRIF	Caribbean Catastrophe Reinsurance Facility
CDP	Carbon Disclosure Project
CEA	Cost-Effectiveness Analysis
CO_2	Carbon Dioxide
CRW	Crisis Response Window
DAMPA	Damayan ng Maralitang Pilipinong Api
DDI	Disaster Deficit Index
DEFRA	UK Department for Environment, Food & Rural Affairs
DFID	Department for International Development
DRM	Disaster Risk Management
EC	European Commission
EIA	Environmental Impact Assessment
EIU	Economist Intelligence Unit
EMS	Environmental Management System
EP	Environmental Profile
ERA	Environmental Risk Assessment
ERM	Environmental Resources Management
ESG	Environmental, Social and Corporate Governance
EU	European Union
EWS	Early Warning System
FEMA	Federal Emergency Management Agency
FONDEN	Fund for Natural Disasters
GAR	Global Assessment Report

GDP	Gross Domestic Product
GDRC	Global Development Research Center
GFDRR	Global Facility for Disaster Reduction and Recovery
GHG	Greenhouse Gas
GSO	Government Office for Science
IADB	Inter-American Development Bank
IBRD	International Bank for Reconstruction and Development
IDA	International Development Association
IFC	International Finance Corporation
IFI	International Financial Institution
IFRC	International Federation of Red Cross and Red Crescent Societies
IIASA	International Institute for Applied Systems Analysis
IMF	International Monetary Fund
IPCC	Intergovernmental Panel on Climate Change
IRM	Immediate Response Mechanism
JaREEACH	Jamaica Rural Economy and Ecosystems Adapted for Climate Change
LIC	Low-Income Country
MCA	Multicriteria Analysis
MDB	Multilateral Development Bank
MIC	Middle-Income Country
ODI	Overseas Development Institute
OECD	Organisation for Economic Co-operation and Development
PDF	Probability Density Function
PDR	People's Democratic Republic
PML	Probable Maximum Loss
RCF	Rapid Credit Facility
RDB	Regional Development Bank
SCF	Standby Credit Facility
SDG	Sustainable Development Goal
SIDS	Small Island Developing State
tCO_2e	Tonnes of CO_2 equivalent
TFP	Total Factor Productivity
UK	United Kingdom
UNDESA	UN Department of Economic and Social Affairs
UNDP	UN Development Programme
UNEP	UN Environment Programme
UNESCAP	UN Economic and Social Commission for Asia and the Pacific
UNISDR	UN Office for Disaster Risk Reduction
US	United States
WEF	World Economic Forum
WFP	World Food Programme
WRI	World Resources Institute

Chapter 1
The Triple Dividend of Resilience—A New Narrative for Disaster Risk Management and Development

Thomas Tanner, Swenja Surminski, Emily Wilkinson, Robert Reid, Jun Rentschler, Sumati Rajput and Emma Lovell

Abstract To secure development gains and help eradicate poverty in the long run, it is critical to strengthen *ex-ante* disaster risk management (DRM) measures that build resilience at the household, firm and macro level. Decision-makers however often view DRM investments as a gamble that pays off only in the event of a disaster. This is despite increasing evidence that building resilience yields significant and tangible benefits, even if a disaster does not happen for many years. This chapter outlines the Triple Dividend of Resilience as a new analytical method to enhance the business case for investments in building resilience. The three benefits that are outlined are: (1) avoiding losses when disasters strike; (2) unlocking development potential by stimulating economic activity thanks to reduced disaster-related investment risks; and (3) social, environmental and economic co-benefits associated with investments. The second and third dividends in particular are typically overlooked in appraisals around investment decisions, and can accrue even in the absence of disaster events.

T. Tanner (✉) · E. Wilkinson · E. Lovell
Overseas Development Institute, London, UK
e-mail: T.tanner@odi.org.uk

E. Wilkinson
e-mail: e.wilkinson@odi.org.uk

E. Lovell
e-mail: e.lovell@odi.org.uk

S. Surminski
Grantham Research Institute on Climate Change and the Environment, London School of Economics, London, UK
e-mail: surminski@lse.ac.uk

R. Reid · J. Rentschler · S. Rajput
Global Facility for Disaster Reduction and Recovery (GFDRR) at the World Bank, Washington, DC, USA
e-mail: rreid1@worldbank.org

J. Rentschler
e-mail: jrentschler@worldbank.org

S. Rajput
e-mail: srajput1@worldbank.org

S. Surminski and T. Tanner (eds.), *Realising the 'Triple Dividend of Resilience'*, Climate Risk Management, Policy and Governance,
DOI 10.1007/978-3-319-40694-7_1

Presenting evidence of additional dividends to policy-makers and investors can provide a stronger case for investment in DRM, helping to reconcile short- and long-term objectives. This chapter sets the conceptual basis for the more detailed assessments of the resilience streams and implications for decision-makers provided in the following chapters.

Keywords Disaster risk management · Resilience · Development · Economic losses · Co-Benefits · Decision making · Triple Dividend of Resilience

1.1 The Case for Investing in Resilience

1.1.1 Disasters, Poverty and Development

There is growing awareness that disaster and climate risk threatens future growth and development. The total number of disaster events has been increasing since the 1980s, with this trend set to continue, driven by climate change, population growth, urbanisation, more people living in coastal areas and floodplains and the degradation or loss of natural ecosystems (Field et al. 2012; UNISDR 2015a). Economic losses from "natural" disasters are now reaching \$150–200 billion each year, up from \$50 billion in the 1980s (see Fig. 1.1), while projected future disaster losses in the built environment alone are estimated at \$314 billion per year (UNISDR 2015a).

The increasing frequency of devastating disasters is a major obstacle to the reduction of poverty and promotion of shared prosperity. While progress in human

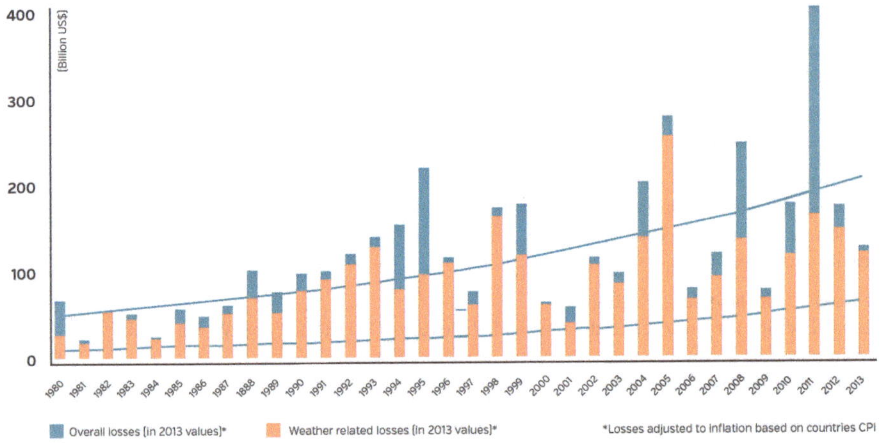

Fig. 1.1 Disaster and weather-related losses, 1980–2013 (\$ billions). *Source* , in GFDRR (2015)

development has been remarkable in the past two decades, with global levels of extreme poverty likely to fall to under 10 % of the global population in 2015 (World Bank 2015), gains have not been evenly distributed between or within countries (World Bank 2013). Without concerted action, by 2030 there could be up to 325 million extremely poor people living in the 49 countries most exposed to natural hazards and climate extremes, the majority in South Asia and Sub-Saharan Africa (Shepherd et al. 2013). Urban growth will be a particularly strong driver, with the global urban population increasing by 1.4 million each week, roughly the size of Stockholm (Global Commission on the Economy and Climate 2015). Most of this expansion is occurring in low- and middle-income countries, where the growth of informal settlements amplifies disaster risk as low-income families are forced to occupy hazard-prone areas with low land values, deficient infrastructure, a lack of social protection and high levels of environmental degradation (UNISDR 2015a).

1.1.2 Incentivising **ex-ante** *Disaster Risk Management*

A variety of reports have emphasised the need to incentivise and enable greater *ex-ante* disaster risk management (DRM) (Field et al. 2012; UNISDR 2015a; World Bank 2013). A range of international policy frameworks echo this message, including the Sendai Framework for Disaster Risk Reduction, the Financing for Development Framework, the Sustainable Development Goals and the climate change agreements, all of which these highlight the importance of investing in resilience. However, it still is not happening at the rate needed to curb rising disaster losses. Although some countries, cities and communities have made progress, funding remains heavily biased towards *ex-post* measures. Meanwhile, the importance of such *ex-ante* prevention is not yet reflected much in the policy and practice of governments, aid agencies, communities or businesses (Kellett and Caravani 2013).

There are many reasons for this underinvestment in disaster resilience. These include lack of resources in poor countries, limited understanding of risks and impacts, greater political buy-in for more visible post-disaster support initiatives and the ready availability of international post-disaster assistance (Keefer 2009; Wilkinson 2012; World Bank 2013). In particular, DRM suffers from a lack of salience with citizens, as the benefits are hard to perceive (Wilkinson 2012). Crucially, policy-makers tend to underinvest or not invest at all in projects to manage risk because the costs of such investments are visible and immediate, whereas their direct benefits and the distribution of these are unclear, uncertain and distant. Existing methods of appraising investment decisions often fail to incentivise DRM because they undervalue the resulting benefit streams.

There are also reasons why individuals choose to stay and invest in risky areas (Chap. 2). Increased exposure to natural hazards may be seen as an unavoidable side-effect of investments to create additional employment and growth from international trade in areas characterised by low transportation costs but exposed to flood risks (Gallup et al. 1998). In China, for instance, total factor productivity (TFP) is 85 % higher in coastal regions than inland, and TFP growth is not significantly different despite higher investment in inland regions, suggesting lower transport costs offer a permanent productivity advantage in coastal regions (Fleisher and Chen 1997). Similarly, poor people living in flood-prone areas in Mumbai are well aware of the risks and make deliberate decisions to live where they do to benefit from higher wages and better schools and medical care (Patankar 2015).

To counter these problems, this chapter examines a shift in the narrative away from a singular focus on losses as a driver for action towards the recognition and appraisal of a broader set of dividends from investing in DRM. We argue that DRM investment should be considered as something that is good for wealth, wellbeing, profit, growth and sustainable development, in addition to preventing human and economic losses should a disaster strike. Through the use of the triple dividend concept, we examine evidence of the wider benefits of investing in resilience measures with the intention of improving awareness and stimulating the development of appraisal tools that can incorporate these factors and enhance future investments in DRM.

1.2 The Triple Dividend: A Comprehensive Business Case for Resilience

Investing in DRM yields a wide range of benefits in the short and long term: if a disaster does strike, then prior planning and investments help reduce human and economic losses. This is the basic rationale and common narrative for DRM, associated with saving lives, reducing losses and supporting both individuals and communities to quickly and effectively bounce back from disasters. However, there is a range of other resilience dividends (Rodin 2014; WRI 2008) associated with DRM investments. The risk of disasters creates background risk, which constrains investment in capital productivity and development for fear of disaster events eroding returns. DRM enables forward-looking planning, long-term capital investments and entrepreneurship. These are all crucial elements for economic growth and shared prosperity. In addition, investments in DRM and resilience generate wider social, economic and environmental co-benefits irrespective of disasters. These could include multiplier effects on employment or trade or strengthening water and sewage systems. Importantly, many investments can be specifically designed to have a dual use, such as roads that act as embankments or tunnels that can also serve as water retention and drainage systems. As such, determining whether an investment is a DRM measure with development co-benefits or a development measure with DRM co-benefits is often a matter of perspective.

This chapter argues that a more complete understanding of this wide range of benefits—or dividends—of DRM investments is critical for strengthening the business case for building resilience. In particular, we propose three concrete dividends from *ex-ante* DRM measures:

1. **The first dividend** (“**avoided losses**”). Investing in DRM strategies takes the form of reduced losses and damages in the event of a disaster. These losses and damages can be both direct and indirect, leading to both immediate and

long-term effects. Most notably, the first dividend includes saved lives, along with prevented or reduced damage to infrastructure and assets. This corresponds to the conventional *ex-post*, loss-centric view, and is likely to underestimate the benefits of DRM measures.

2. **The second dividend** (“**unlocking economic potential**”). Even the mere possibility of a future disaster has real impacts on present-day economic growth, particularly in regions or localities where disaster risks are perceived to be high. DRM measures help manage this ever-present background risk of potential future disasters. This helps unlock economic development potential by enabling forward-looking planning and investment. Increased resilience can catalyse innovation, entrepreneurship and investment in productive assets—even if disasters do not occur for a long time.
3. **The third dividend** (“**generating development co-benefits**”). DRM investments are typically associated with economic, social and environmental uses, or “co-benefits”. Co-benefits can play an important role in motivating DRM measures and determining their design (e.g. shelters doubling as community spaces or flood protection infrastructure doubling as roads). While the nature of co-benefits varies significantly, they all materialise even in the absence of a disaster.

Figure 1.2 summarises the three dividends of resilience. This chapter is a first step in bringing together evidence that helps characterise the dividends resulting from DRM investments. These are used to build the case for an incentive structure for DRM that goes beyond avoided losses.

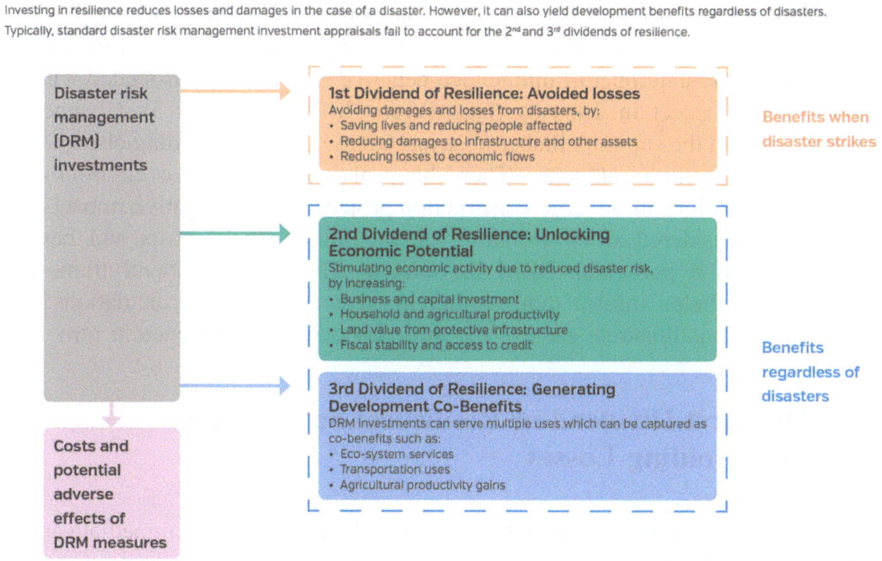

Fig. 1.2 The Triple Dividend of Resilience—strengthening the case for investing in DRM

For example, the World Bank and Mexico's Ministry of Finance elaborated a joint study to determine the impact of investment in flood defence in terms of reducing flood damage in the state of Tabasco between 2007 and 2010. The first dividend was revealed by the cost–benefit ratio of these benefits, which was 4:1, contributing to avoided damages and losses when floods occurred in 2010 equivalent to $3 billion, or 7 % of the gross domestic product (GDP) of Tabasco (World Bank 2014a). This high ratio supports the business case of investing in DRM, but it could be further strengthened if it captured the full range of dividends associated with these investments, such as new flood defences helping maintain existing and stimulate new investments. For instance, continued investment by companies in the Tabasco region that were otherwise leaving further strengthens the case for avoided losses. Additionally, the second dividend is evident from reduced background risk encouraging private investment in housing in previously flood-prone areas and public investment in improved drainage and electricity networks in areas where floods had previously deterred such investment (ibid.).

In addition, the capital of Tabasco, Villahermosa, has seen improvements in the urban environment as a result of federal government investment in flood defence. Major DRM investments have stimulated local actors to take greater care of the environment while small-scale projects with environmental benefits have been initiated, including tree-planting on riverbanks to prevent landslides, which could potentially reveal environmental co-benefits as mentioned under the third dividend. People are beginning to dispose of litter more responsibly, throwing less on the streets or into drains, helping avoid blockages during the rainy season (see Chap. 3).

This example also demonstrates the need to examine the possible negative consequences, which could be considered negative co-benefits associated with a comprehensive assessment. For example, a report by the Colegio de la Frontera Sur suggests there are a number of unintended negative externalities associated with the flood defence project in Tabasco (Díaz-Perera 2013; see Chap. 3). Channelling water away from the capital Villahermosa has led to increased flooding elsewhere in the state, mainly in rural areas. There have also been negative environmental impacts as a result of these large construction projects. These negative impacts also need to be considered when weighing up the full range of costs and benefits associated with a particular DRM investment. The triple dividend framework presented here helps inform more comprehensive cost–benefit calculations. The following sections illustrate each of the three dividends of resilience in turn.

1.3 The First Dividend of Resilience: Saving Lives and Avoiding Losses

DRM measures can avoid or reduce losses and damages (both immediate and long run) in the event of a disaster. They include:

- Saving lives and reducing numbers of people affected
- Reducing direct damages to infrastructure and other assets

- Reducing economic and non-monetary losses (direct and indirect).

The Triple Dividend of Resilience approach is motivated by the observation that fully acknowledging the benefits of resilience will strengthen the business case for DRM investments. However, while other benefits of DRM can play substantial roles, the primary objective of DRM remains clear: to save lives, while also reducing loss and damage to people and their assets. In recognition of the importance of this objective, this section briefly highlights the evidence for effective risk management that limits human and economic disaster losses.

1.3.1 Saving Lives and Reducing Number of people Affected

Effective DRM policies and actions are often measured by their ability to save lives and reduce the number of people affected by disasters. To this effect, progress in saving lives has been marked. As reported in the 2015 Global Assessment Report (GAR), "improvements in disaster management have led to dramatic reductions in mortality in some countries" (UNISDR 2015a). In Bangladesh, deaths from cyclones have been reduced considerably, owing to a combination of strengthened coastal defences, cyclone shelters and early warning systems (EWS).

While comparisons across countries and events are difficult because of contextual differences, it is possible to infer levels of preparedness and effectiveness of DRM measures through observing the impacts of similar hazards (see Fig. 1.3). In 2010, the existence and enforcement of building codes helped limit earthquake damage in Chile, with less than 1000 people killed, despite a magnitude 500 times greater than the Haiti quake of the same year that killed over 230,000 (Lovett 2010). More recently, increased investment in infrastructure and disaster preparedness paid off in the latest earthquake and tsunami in September 2015 in Chile, which resulted in relatively low casualties, despite a 8.2 magnitude (UNISDR 2015b). Volcano-related mortality has also decreased significantly as a result of volcano monitoring, assessments and EWS; and, although not all volcanoes are monitored, it is estimated that such measures have saved about 50,000 lives over the past century (Auker et al. 2013).

DRM interventions can also save lives through acknowledging different people's needs, vulnerabilities and capacities. Integrating indigenous knowledge into DRM initiatives has been shown to help avoid loss of life. For example, oral history on ocean and buffalo behaviour meant the inhabitants of Simeulue Island in Indonesia had early warning before the Indian Ocean tsunami in 2004 and were able to retreat to the hills. As a result, only seven out of 78,000 people died in the tsunami, despite

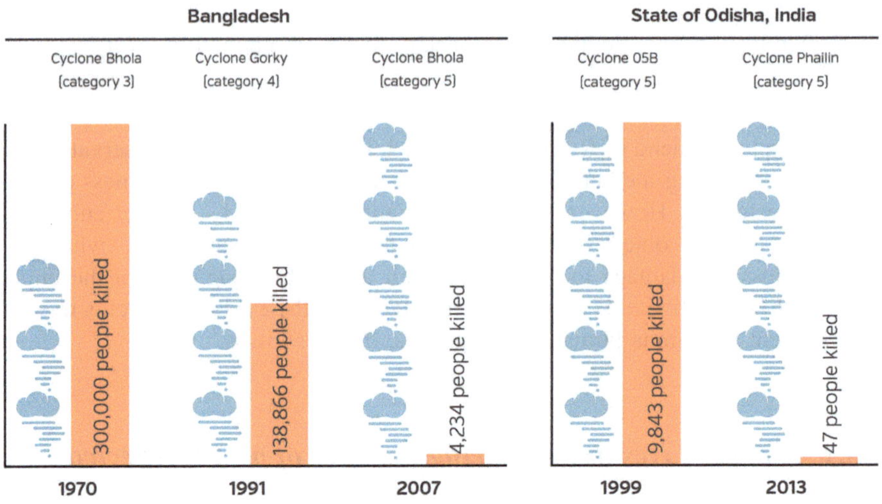

Fig. 1.3 Reduced cyclone mortality in Bangladesh and Odisha, India. *Source* Munich Re cited in GFDRR (2015)

the island being located just 40 km from the epicentre of the earthquake (Lovell and le Masson 2014).

1.3.2 Reducing Damages and Losses

There is a strong body of evidence for the effectiveness of DRM measures gathered from projects around the world. The 2015 GAR concludes that "annual global investment of US$6 billion in appropriate disaster risk management strategies would generate total benefits in terms of risk reduction of US$360 billion. This is equivalent to an annual reduction of new and additional average annual loss by more than 20 %" (UNISDR 2015a). Mechler and Bouwer (2015) make the case that, despite the increase in risk exposure, various DRM strategies have decreased vulnerabilities throughout the world.

Infrastructure, EWS and planning are three areas where DRM investments have been critical in reducing losses from disasters. Infrastructure losses have particularly profound consequences for development progress. Between 2015 and 2030, approximately $90 trillion is expected to be invested globally in infrastructure to meet the world's urban, land use and energy needs (Global Commission on the Economy and Climate 2014). This is particularly pertinent in Asia and Africa, where 90 % of urban growth is expected to take place between now and 2050, which will result in accompanying infrastructure needs (UNDESA 2014). It is crucial that these huge financial investments are disaster-resilient, as this will protect lives and secure development progress.

Infrastructure losses often go well beyond physical damage. Business losses can be the consequence of "ripple effects" as the impacts of shocks propagate both upstream (backward) from clients to suppliers and downstream (forward) from suppliers to clients. The 2011 Thai floods' impact on global supply chains forced Toyota to slow down production in factories in Indonesia, Japan, Malaysia, Pakistan, the Philippines, South Africa, Vietnam and North America. Locating industrial parks in protected areas less prone to flooding would have improved disaster resilience and reduced losses (Scor SE 2013). Similarly, the Tohoku-Pacific earthquake in Japan in March 2011 reduced domestic industrial production and the exports of goods used as inputs in the auto industry, leading to a reported cut in production at Toyota's Indian subsidiary by up to 70 % between 25 April and 4 June (The Economic Times 2011).

EWS are frequently cited for their role in reducing economic losses of disasters by triggering other important prevention actions, as there is more lead time to protect assets. While issues of attribution and lack of widespread cost benefit calculations complicate the evidence base (Rogers and Tsirkunov 2011), Table 1.1 suggests significant loss and damage reduction is possible owing to an early warning of different lead times on a number of different movable assets (Subbiah et al. 2008).

Establishing and enforcing risk-informed, locally appropriate standards and codes for new buildings and other infrastructure reduces the risk of damage to structures in the event of a disaster. Existing infrastructure can also be retrofitted to adhere to building standards. For example, Cyclone Ian in Tonga in 2014 had significantly less impact on houses constructed to cyclone standards in the early 1980s than it did on many newer houses that were not built in compliance with the standard. These were completely destroyed or severely damaged (GFDRR 2014). Similarly, homes built with typhoon-resistant features as part of the Storm Resistant Housing for a Resilient Da Nang City project in Vietnam showed no damage when Typhoon Nari hit in October 2013 (Tran 2013).

A World Bank study of earthquake vulnerability in Colombia (Ghesquiere et al. 2006) assessed a range of measures that were:

- Structural (retrofitting and reinforcement of public buildings, such as schools, hospitals, fire stations and administrative buildings)
- Non-structural (the resettlement of vulnerable populations in high-risk areas)
- Functional (protection of people and assets, so they remain functional during and immediately after an emergency).

A probabilistic cost–benefit analysis then helped demonstrate to decision-makers the significant reductions in probable maximum loss (PML) of a one-in-1000-year earthquake event, before and after structural investments were made (shown in Fig. 1.4). The average annual returns on mitigation investments for schools, hospitals and fire stations were estimated to be as high as 19 % for structural investments and 32 % for structural and functional investments. In addition to the direct costs of structural and functional assets, there may be significant indirect losses. One

Table 1.1 Damage reduction owing to early warning of different lead times

Item	Lead time	Damage reduction (%)	Actions taken to reduce damages
Household items	24 h	20	Removal of some household items
	48 h	80	Removal of additional possessions
	Up to 7 days	90	Removal of all possible possessions including stored crops
Livestock	24 h	10	Poultry moved to safety
	48 h	40	Poultry, farm animals moved to safety
	Up to 7 days	45	Poultry, farm animals, forages, straw moved to safety
Agriculture	24 h	10	Agricultural implements and equipment removed
	48 h	30	Nurseries, seed beds saved, 50 % of crop harvested, agricultural implements and equipment removed
	Up to 7 days	70	Nurseries, seed beds saved, fruit trees harvested, 100 % of crop harvested, agricultural implements and equipment removed
Fisheries	24 h	30	Some fish, shrimps, prawns harvested
	48 h	40	Some fish, shrimps, prawns harvested, nets erected
	Up to 7 days	70	All fish, shrimps, prawns harvested, nets erected, equipment removed
Open sea fishing	24 h	10	Fishing net, boat damage avoided
	48 h	15	Fishing nets removed, boat damage avoided
School or office	24 h	5	Money, some office equipment saved
	48 h	10	Money, most office equipment saved
	Up to 7 days	15	Money, all office equipment, including furniture protected

Source Subbiah et al. (2008)

example of this would be the way disruption to education can constrain future career options and prosperity later in life.

The examples above suggest the avoidance of loss usually provides a critically important stream of benefits for DRM investments. Widening avoided loss calculations beyond immediate asset losses to include the impact of disasters on the wider economy and society can help strengthen the case for investing in DRM. Nevertheless, several factors limit the potential incentivising role. First, these wider benefits are hard to identify, calculate and attribute. This is because they rely on counter-factual reasoning, in that a DRM investment will reduce the probability of a disaster happening. It is difficult to measure the impact of something not happening. Second, those actors facing the costs of investment may not enjoy the benefits, although this may not be a problem for governments concerned with wider economic and social goods. However, the most critical point to consider here, from the perspective of this chapter, is that using loss-based approaches to justify investment is reliant on the occurrence of a disaster event in the future, which is a major flaw. By identifying the dividends of resilience that are delivered even in the absence of

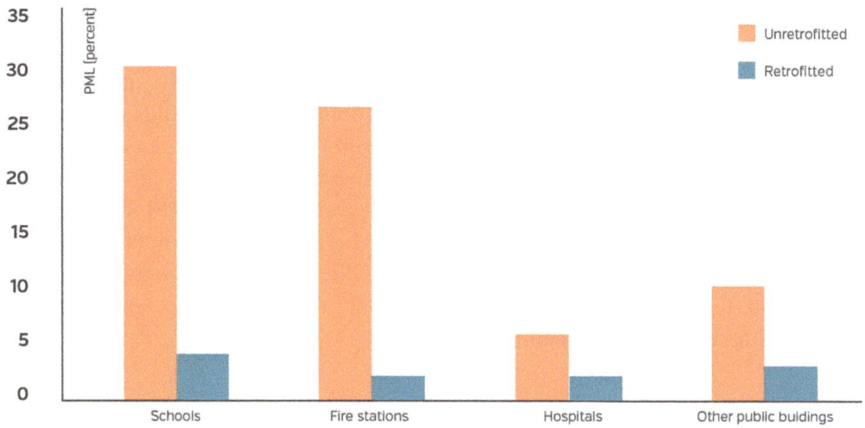

Fig. 1.4 PML of a 1000 year earthquake event, before and after structural investments. *Source* Ghesquiere et al. (2006)

disaster events, and incorporating them in decision-making, the case for investing in resilience can be greatly improved.

1.4 The Second Dividend of Resilience: Unlocking Economic Potential

DRM measures that reduce the background risk resulting from potential future disasters can have immediate and significant development benefits. Increased resilience enables forward-looking planning, long-term capital investments and entrepreneurship, even if disasters do not occur for a long time. These benefits include:

- Economic gains from positive risk-taking (e.g. entrepreneurship and innovation)
- Investments in productive assets (e.g. in small-scale agriculture)
- Extending planning horizons (e.g. for building up savings)
- Increase in land values after DRM investment.

In disaster-prone places, risks of extreme weather events and disasters create an ever-present background risk. As a consequence, risk-averse households and firms avoid long-term investments in productive assets, entrepreneurship is restricted and planning horizons are shortened, meaning development opportunities are lost. By reducing this background risk, or by helping households and firms manage it effectively, DRM measures can have immediate and significant economic benefits.

This section presents evidence that investments in *ex-ante* DRM can unlock economic opportunities for households, government and the private sector, as well as more broadly, at the macroeconomic level. And, more broadly, at the macroeconomic level. For example, the evidence from poor rural households dependent on agricultural income suggests strengthening *ex-ante* DRM enables those households to increase savings and investment in productive assets, thereby improving their productivity and livelihoods. Further examples show how DRM measures can increase land values, as well as improve credit access, fiscal management and public sector coordination. Overall, increased resilience can be seen as a catalyst for positive risk-taking such as capital investments, entrepreneurship and innovation, along with forward-looking planning.

1.4.1 Increased Business and Capital Investment

Without effective instruments for managing disaster risks and the adverse consequences of disasters, investment decisions are likely to be excessively risk-averse (Elbers et al. 2007; Gollier and Pratt 1996). As a result, businesses refrain from engaging in entrepreneurial activities and innovation or making long-term investments in productive assets.

Therefore one of the most immediate benefits that investing in DRM has to offer the private sector relates to investment risk-taking. Taking positive risks, engaging in entrepreneurial activities and investing in productive assets and innovation are the drivers of job creation, rising incomes, greater productivity and overall economic growth. However, the perceived risk of future disasters can lead to greater risk aversion, which dampens entrepreneurial activity (Chap. 5). Investing in DRM can help reduce this background risk and provide better information on residual risk, which in turn helps promote the entrepreneurship and investment needed for economic growth and job creation. While risk-taking can increase welfare, there may be a trade-off between exposure to natural hazards and productivity or economic growth in high disaster risk situations. Public and private investment in improved risk management can mitigate this trade-off, reducing the background risk that prevents people from investing, therefore improving productivity and accelerating growth (Hallegatte 2014).

Similarly, disaster insurance can encourage the kind of "positive risk-taking" that is arguably fundamental to the development process, making investments more secure and therefore fostering business innovation and growth (Chap. 2). However, disaster insurance may also lead to moral hazards if it is not designed with the correct control measures in place. This points to a potential counter effect of using insurance, where it can create a false sense of security, increase vulnerability to exceptional events or encourage inappropriate development in high-risk areas (Surminski 2014).

Furthermore, investing in DRM can generate benefits that extend across sectors to the macroeconomic level. A region- or country-wide boost to productive investments can boost the overall development of a country. Protecting coastal regions, towns, business districts or ports from flood can foster economic activity, long-term planning and capital investments. Well designed and maintained DRM investment, large DRM infrastructure investments (such as dikes) can protect not only large firms themselves but also their workers and suppliers, along with their social and logistic infrastructure. If DRM investments enable firms and their stakeholders to make long-term capital investments, engage in trade and thus promote business development, the entire geographic area benefits collectively (Hallegatte 2014; World Bank 2013).

Firms may also benefit from improvements to their image and credit ratings, through increased stability (Chap. 5). There is some evidence of businesses taking this "good citizen" image seriously; for example, in a set of six case studies of companies describing their activities related to managing the physical impacts of extreme weather and climate change, most saw avoidance of disaster impacts (both now and in the future) as only part of the logic for investing in resilience (Crawford and Seidel 2013). Companies such as American Water, The Hartford, National Grid and Rio Tinto all emphasised that fulfilling, or staying ahead of, regulatory and disclosure requirements and new government policy were key business drivers for investing in resilience. A survey of European companies also revealed that investing in resilience could help develop market opportunities, with 43 % of the companies surveyed anticipating increased demand for existing products/services (CDP 2015).

1.4.2 Household and Agricultural Productivity Dividends

When levels of background risk are high, evidence suggests households lacking effective risk management tools will tend to spread their overall risk. Rather than specialising, households tend to engage in a wider range of lower-risk activities, thereby reducing returns to assets and investments (Chap. 2). For example, there is evidence that rural households avoid focusing solely on agriculture and instead diversify occupations within households as a risk management measure—with negative impacts on long-term welfare (Rentschler 2013). While such actions reduce the risk of severe losses, they obstruct growth and incentives to invest (Carter and Barrett 2006; Dercon 2005).

An illustration of this effect in an agricultural context can be found in Zimbabwe. Here, farmers exposed to risk exhibit a mean capital stock that is half as large as for farmers who are not exposed. Of this reduction in capital, *ex-ante* risk accounts for two thirds of the difference; hence, most of the welfare impact of risk comes through reduced investments and risk-taking, not damage and loss when a disaster occurs (Elbers et al. 2007; Chap. 2). Extending these findings into other decision-making contexts and sectors could provide crucial evidence to enhance the incentives for *ex-ante* investments in DRM.

Household insurance and social safety net programmes have been observed to stimulate savings, investment in productive assets and increases in agricultural productivity in a number of different countries, with subsequent improvements in income levels. In Ethiopia, the R4 Rural Resilience Initiative (previously the Horn of Africa Risk Transfer for Adaptation programme) is providing risk management support, including weather-indexed insurance to small-scale and subsistence farmers. Premiums are largely paid through labour to support risk management activities. In the event that rainfall drops below a predetermined threshold during the growing season, insurance payments are automatically triggered.

An evaluation of the programme has found that insurance is enabling farmers to increase their savings, which can act as an important reserve in the case of contingencies. Moreover, insured farmers have been found to increase their investments in productive assets, in particular oxen, but also fertiliser, improved seeds and compost—thus improving their overall productivity (Greatrex et al. 2015; Madajewicz et al. 2013).

Evaluations of the Mexican government's Committee for Natural Disasters and Emergencies programme show how weather-indexed insurance not only helps compensate for drought losses but also directly increases the productivity of small-scale farmers. The insurance programme has enabled farmers to overcome credit constraints and mitigated previously chronic underinvestment in tools and fertiliser. As a result, farmers have been able to increase their agricultural productivity, with an average 6 % increase in maize yields. Evidence also shows that insured farmers invest in riskier but higher-yielding cultivation methods, with higher overall planting-stage investments than uninsured peers, enabling them to reconcile entrepreneurial investment decisions with effective risk management (Dar et al. 2013; Emerick et al. 2015).

Overall, these evaluations demonstrate how effective risk management tools not only yield significant benefits in the aftermath of a disaster but also can yield significant benefits even if disasters do not strike for many years, such as through increases in productivity and income levels. By reducing background risk, DRM measures can directly influence economic decisions and behaviour, actively contributing to a long-term sustainable economic development process. If implemented at sufficient scale, DRM measures (such as weather-indexed insurance programmes) can have significant economic development benefits at the macro level, and even be cost-effective in the absence of disasters.

1.4.3 Land Value Dividends from Protective Infrastructure

Investment in dams, levees and other structures to protect assets from disaster impacts can unlock economic potential through increases in productive investment and consequent increases in the value of land. To some extent, the efficiency of infrastructure provision can be measured by the relationship between land value capitalisation and infrastructure costs. When the benefits of capitalised land values exceed the costs of installing infrastructure, infrastructure is generally undersupplied. This relationship can be seen in Table 1.2, which demonstrates land value gains and infrastructure costs in Recife, Brazil. In this case, there is clearly a need for more investment in road pavement and wastewater removal in order to meet economic demand, as the land value gains exceed the costs of infrastructure supply. This is in contrast with the water supply, which has an almost equal land value capitalisation to investment cost ratio of 1:1 (Peterson 2012).

In a similar way, protective infrastructure can also generate dividends of resilience. Hard infrastructure for protection, along with soft DRM measures, such as monitoring and early warning, can protect assets from disaster impacts. These factors are likely to have a positive effect on land prices, which also shows an increased willingness for people to invest in these areas, given a reduced background risk. These increased land values can in turn help raise government revenue, helping finance the cost of *ex-ante* DRM measures. It is possible to learn from building development projects, where one of the most common strategies for recovering infrastructure costs involves the sale of land with enhanced value. Here, the business case for protective infrastructure investments can be more accurately costed, particularly where the public sector owns the land.

1.4.4 Fiscal Stability and Future Credit Risks

There are a number of economic and other benefits of DRM to be recognised and realised by those in charge of fiscal policy decisions. Approaches organised around

Table 1.2 Land value gains and infrastructure costs in Recife, Brazil

	Increase in land value ($ per square meter) by distance to centre			
Service	5–10 km	15–20 km	25–30 km	Ratio of gain in land value to investment cost
Water supply	11.1	5.1	3.2	1.02
Road pavement	9.1	4.8	3.4	2.58
Wastewater removal	8.5	1.8	0.3	3.03

Source Peterson (2012)

the protection of the balance sheet using risk-financing instruments have seen growing emphasis in disaster-prone countries in recent years (Chap. 4). The inclusion of disaster risk in these instruments and shock-financing mechanisms can have a significant impact on reducing uncertainty, potentially unlocking higher private investment, employment and growth (Chap. 7). Implementing a structured process for risk detection in the balance sheet can potentially provide a "price signal"; terms. This is the case in Mexico, where innovative financing arrangements have been initiated under National Disaster Fund to incentivise investment, to "build back better" and relocate housing to lower-risk areas (Hoflinger et al. 2012). In contrast, a focus on ex -post disaster management offers little in the way of risk awareness or stimulating risk reduction (Phaup and Kirschner 2010).

One example of a strategic DRM response that incorporates the triple dividend concept is the fiscal risk matrix. Such matrices combine the assessment of many different contingent risks, including their interaction with disaster risk, and their use has grown from insights gained during recent financial and fiscal crises (Chaps. 4 and 7). Fiscal risks are "stress-tested" through sensitivity tests on baseline macro and fiscal indicators. There is also a growing understanding of the need to take a systematic perspective in understanding the potential for complex and interrelated shocks, leading to a multi-risk approach (WEF 2015). Disaster risk has come to be considered a key threat; in a recent survey regarding relevant fiscal risks in Organisation for Economic Co-operation and Development countries, disasters emerged as an important concern (Kopits 2014).

In the future, the benefits of lower background risk may also be reflected in businesses and governments' access to affordable credit. Noting the growing influence of climate change on risks, Standard and Poor's suggest climate change could feed through to sovereign creditworthiness through economic, fiscal and external performance (Standard & Poor's Rating Services 2014). Credit rating agencies have also recognised that companies' credit profiles may be determined to a larger degree in the future by climate-related disasters and the increased exposure of companies and their global supply chains to risk (Moody's 2015; Standard & Poor's Rating Services 2015). In some cases, credit rating agencies have explicitly called for DRM strategies to both prevent disaster losses and maintain credit ratings, illustrated in coastal cities in south-eastern Virginia's Hampton Roads region of the US (Moody's 2015). Access to credit to enable capital investment may therefore provide a component of the development dividend for firms, with ratings agencies now calling for greater disclosure of firms' exposure to extreme natural hazards, which should encourage them to bolster their resilience to these events and aid transparency (Standard & Poor's Rating Services 2015).

1.5 The Third Dividend of Resilience: Co-benefits of DRM Investments

DRM investments have multiple uses, which can be classified as economic, social and environmental co-benefits. These co-benefits may be either explicitly designed into the investment (such as dual-use infrastructure) or incidental.

While the nature of co-benefits varies significantly, they all materialise even in the absence of a disaster. Co-benefits can play an important role in motivating DRM measures and determining their design. Multi-purpose design that intentionally integrates these co-benefits can save money and significantly improve the attractiveness of investing in DRM.

These co-benefits include:

- Economic co-benefits (e.g. flood protection supporting fisheries)
- Social co-benefits (e.g. improved transparency or social cohesion)
- Environmental co-benefits (e.g. watershed protection)
- Economic co-benefits (e.g. flood protection supporting fisheries)
- Social co-benefits (e.g. improved transparency or social cohesion)
- Environmental co-benefits (e.g. watershed protection).

To gain a complete picture of the benefits of DRM investments, we must take into account their social, environmental and economic contexts. This makes it evident that DRM measures can yield a variety of co-benefits. These can materialise even in the absence of a disaster, but—unlike the second dividend of resilience—are not because of reduced background risk. In line with growing efforts to highlight the co-benefits of climate change mitigation, it is critical for decision-makers to fully understand and account for the co-benefits of DRM and climate change adaptation measures (Chap. 3; Global Commission on the Economy and Climate 2014; Kok et al. 2008; Tanner et al. 2015; Santucci et al. 2015). Some of these might be unintentional and generated as "spill-over" effects. As emphasised above, it is important that the design of DRM measures also fully consider and mitigate the potential negative side-effects of DRM measures (such as the costs of relocation of communities from risky areas).

However, the examples below also demonstrate the diverse synergies that can be created by intentionally designing measures to deliver both DRM and development objectives. Conversely, linking with DRM goals can also help deliver other benefits that might otherwise be undersupplied, such as public green space or improved transport networks.

Multi-use design is becoming increasingly common in physical DRM infrastructure, where high upfront costs might otherwise make the investments harder to justify. Cyclone shelters in Bangladesh have a long history of multi-purpose design for use outside storm times (Khan 2008). In Tinputz district, Papua New Guinea, resilient infrastructure for education and health is designed both as a space for communal gatherings and as safe shelters for the community if disaster does strike (Tinputz District Disaster Risk Management Committee 2014).

Table 1.3 presents examples of the breadth of these co-benefits, illustrating how widely they can vary in practice. Some co-benefits can be directly observed, measured and quantified, such as livelihood benefits or dual purpose infrastructure; others, such as social cohesion, can be very hard to quantify and integrate in economic analyses, despite being potentially significant. Below, we outline three areas where DRM activities are delivering co-benefits: ecosystem-based approaches, transport systems and agricultural projects.

1.5.1 Ecosystem-Based Co-benefits

Ecosystem-based approaches to DRM and climate adaptation help to illustrate the co-benefits from investing in resilience. These have gained popularity in recent years, emphasising how good stewardship of environmental systems can help reduce and adapt to disaster risks, in turn saving lives and reducing loss and

Table 1.3 The range of co-benefits associated with DRM measures

DRM activity	Possible co-benefits
Flood protection structures	Provision of irrigation or potable water and hydro-electric power Dual-purpose road infrastructure
Strengthening DRM capacity of civil society	Improved governance, more organised social structures
Ecosystem-based DRM approaches	Environmental conservation, improved air quality, climate change mitigation
Shelters	Community facilities (e.g. clinics or schools) in non-disaster periods
Improving water supply systems in rural areas	Water supply systems improved regardless of a disaster occurring
Construction and use of drainage pipes, canals and water retention basins	Improved irrigation practices, possibly improved agricultural practices Dual purpose road tunnel or parking lot infrastructure
Community-based disaster preparedness	Improved women's involvement in community-level activities
Installing more resilient wireless communications	Enhanced access to telephony and electronic data services
Training farmers to diversify the use of crops	Reduced vulnerability to poverty
Better monitoring of food supplies	Improvement to the food supply chain, possibly making it more cost-effective

Source Adapted from ERM and DFID (2005)

damage. At the same time, ecosystem protection can generate wider social and environmental co-benefits, even in the absence of disaster events. These include:

- Biodiversity conservation
- Carbon sequestration and mitigation
- Land erosion and degradation prevention
- Habitat creation and restoration
- Mitigation of microclimate variability.

Social co-benefits include:

- Improved and secure livelihoods
- Social cohesion and community
- New or preserved recreation areas
- Better quality land for agriculture/livestock
- Better water security.

Aside from economic damages, these approaches have been shown to help develop new or improved income, profits or savings, when compared with alternative DRM or climate adaptation approaches (Doswald et al. 2014). The services delivered by ecosystems do therefore not only offer disaster risk reduction benefits such as flood regulation and protection from storm-surge protection but also enhance food security, provide sustainable water supplies or enhance livelihoods through increasing resource-use options or tourism (Jones et al. 2012; Tanner et al. 2015).

A Vietnam mangrove plantation and DRM project in the typhoon- and flood-prone coastal provinces of northern Vietnam has proven to have significant environmental co-benefits (IFRC 2012), including carbon sequestration, nutrient retention, sediment retention, biodiversity habitat, flood attenuation, wastewater treatment and water supply and recharge. The 17-year-long project cost $8.88 million to set up and has involved the creation of 9462 ha of forest (8961 ha of mangroves) in 166 communes and the "protection of approximately 100 km of dyke lines". The project aims to reach approximately 350,000 beneficiaries directly and 2 million indirectly. There has been an "increase in per hectare yield of aqua culture products such as shells and oyster by 209–789 %". Economic benefits from aqua product collection and honeybee farming are found to be between $344,000 and $6.7 million in the selected communes. Environmental benefits include $218 million alone generated as an estimated minimum of CO_2 emissions absorbed by the planted mangroves (assuming a price of $20/t CO_2e).

Such multi-purpose water management approaches can therefore be designed to provide livelihood, environment, aesthetic or recreational co-benefits alongside disaster resilience. The Netherlands' Room for the River is being designed to manage higher water levels, giving the country's rivers more space to flood safely. The measures also attempt to improve the quality of the immediate surroundings, such as providing new river islands. While in some cases such co-benefits can be assumed to represent good project design and implementation, they are not always costed into the business cases that justify the financing decisions.

Similarly, the World Bank's flood management programme in Sri Lanka's capital Colombo demonstrates the wider value of wetland protection and restoration beyond just flood defence. While performing a valuable role in reducing flood risks, the wetlands of the Colombo basin serve a range of other purposes. They provide livelihoods and economic security to local residents through fishing and rice cultivation, while also serving as a park area for tourism and recreation, and the wetlands and surrounding areas are on average 10 °C cooler than non-pervious areas (such as parking zones or the streets) at the hottest time of the day. This results in energy savings for buildings and homes using artificial cooling systems. Other wetland co-benefits include waste water treatment, maintenance of freshwater supplies, carbon sequestration, climate regulation, water regulation, soil erosion regulation, pollination, recreation and nutrient cycling. Economic analyses of selected wetland co-benefits, including flood protection, carbon sequestration, climate regulation through reduced use of air conditioning near wetland areas and waste water treatment, along with potential income from recreational activities, could be worth \$113–127 million annually (World Bank 2015).

1.5.2 Transport Co-benefits

DRM investments can also be linked with transport systems to combine objectives and improve efficiency. Flood embankments are often used not only to protect the landward assets from inundation but also to support road networks. In doing so, the roads themselves are also more resilient to flood impacts and can permit movement after major hazard events. These synergies can operate at a variety of scales: levees in Bangladesh commonly support small-scale tracks for rickshaws and motorcycles, whereas the 11 dams that protect St Petersburg in Russia against storm surges are built to support 25.4 km of six-lane highway.

The Smart Tunnel scheme in Kuala Lumpur combines storm water flood drainage with vehicle tunnels under the city (see Fig. 1.5). For Category 2 storms, which occur about 10 times each year, part of the flood waters are diverted through the lower section of the road tunnel. For Category 3 storms, which occur once or twice a year, traffic is prohibited and a large part of the flood flow is diverted through the tunnel. A flood detection system provides adequate warning time to evacuate traffic and operate tunnel floodgates as well as to minimise the cost of traffic disruption (Seang 2009).

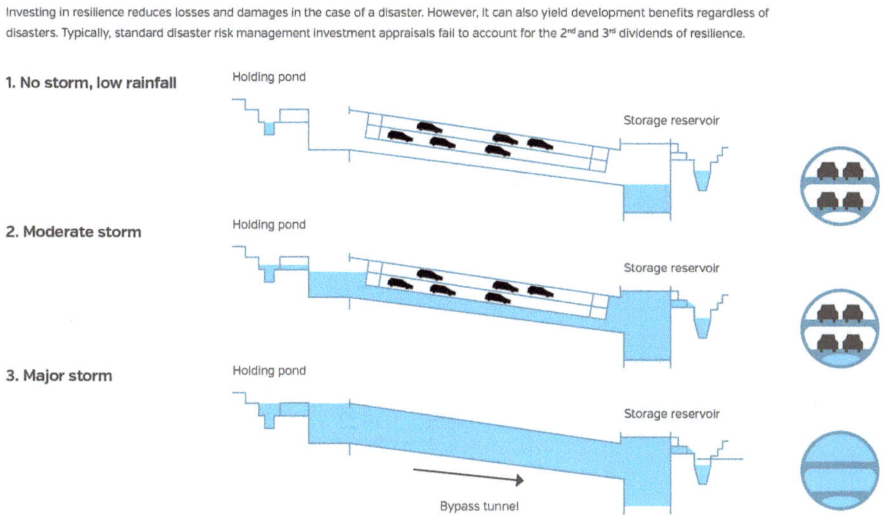

Fig. 1.5 SMART tunnel design in Malaysia

1.5.3 Agricultural Co-benefits

The development of safe sea port shelters as part of the Natural Disaster Risk Management Project in Vietnam were planned to support the sustainable development of the fishing industry. The facilities are highly effective in preventing storm damages for the fishery boats, but also provide a centre for the development of fisheries logistic services. As well as fewer risks to boats related to storms, fisheries business now have a more adequate infrastructure, electricity, water, transportation and other logistics services for their activity (World Bank 2014b).

The World Bank Mekong Integrated Water Resources Management Project has produced a number of economic co-benefits. Estimates of these are based on the assumption that they would be fully reached in three years and that the economic life of the investment would be 15 years. The project was designed with water resource management and flood plain management at its core and resulted in the rehabilitation of 10 floodgates in the Xebangfai River and about 40 village irrigation schemes being put in place in the Xebangfai and Xebanghieng rivers of Lao PDR.

The floodgate rehabilitation increased flood protection (avoiding losses associated with the first dividend of resilience) for 640 ha of cultivated areas, and, on average, $13,200 of flood protection benefits per gate. In addition, co-benefits included increased fish catch in the floodplain, with the average annual benefit of the increased fish catch estimated at $3600 per gate, not only because of reduced

Table 1.4 Summary of floodgate rehabilitation activities

	Required works	Estimated financial cost ($)	Estimated economic cost ($)	Benefit area (ha)	Estimated flood protection benefit ($)	Estimated fish benefit ($)	Total benefit ($)
Huay Pin	Rehabilitation of the mechanical works (gates), minor structural repairs to the headworks	72,000	68,400	120	12,375	3600	15,975
Huay Kae	Rehabilitation of the mechanical works (gates), minor structural repairs to the headworks	52,500	49,875	100	10,313	3600	13,913
Huay Pa Pak	Rehabilitation of the mechanical works (gates), minor structural repairs to the headworks	35,000	33,250	100	10,313	3600	13,913
Huay Bung Or	Rehabilitation of the mechanical works (gates), resectioning of the canal (2.5 km)	31,875	30,281	150	15,469	3600	19,069
Huay Daeng	Rehabilitation of the mechanical works (gates), resectioning of the canal (3.0 km)	38,250	36,338	170	17,531	3600	21,131
Total			218,144	640	66,000	18,000	84,000

Source World Bank (2012)

flooding but also because of enhanced water regulation throughout the year (see Table 1.4). Increases in water use efficiency also produced co-benefits in the form of decreased electricity costs of $2/ha (World Bank 2012).

In Jamaica, the agriculture sector contributes about 6 % of GDP and employs 17–18 % of the labour force. Domestic agriculture is largely located on hillside plots, with an average size of 1 acre with slopes above 15°; meanwhile, the export agriculture (including coffee, banana, cacao and coconut) contributes to 22 % of total exports, raising $274 million in foreign exchange each year. A number of DRM programmes have focused on this sector, including the Jamaica Rural Economy and Ecosystems Adapted for Climate Change (JaREEACH) programme, which aims to strengthen local and national institutional capacity to support climate change adaptation and DRM within agriculture. The Planning Institute of Jamaica has also committed $9.9 million to the development and implementation of adaptation measures, focusing on strengthening agricultural productivity, coastal protection and building local capacity for natural resource management.

Of these investments, those that have focused on reducing drought risk in farming seem to offer particularly high potential for co-benefits. The installation of dedicated irrigation systems to overcome the impact of drought has helped farmers increase their productivity and output as well as reduce soil erosion and deforestation by optimising previously inefficient farming practices (see Box 1.1).

Box 1.1: Key co-benefits of integrated DRM investments in Jamaican agriculture

Economic co-benefits: DRM irrigation projects helped reduce the economic impacts of droughts, particularly in Southern Clarendon and St Elizabeth. These farming communities have also benefited from increased productivity and output relative to other areas, even in the face of drought over the April–June quarter in 2014.

Social co-benefits: Training and shared learning on drip irrigation have strengthened social capital and built comradeship within the communities, especially among the farmers in the field.

Environmental co-benefits: A rainwater catchment tank and drip irrigation system in Lititz, St. Elizabeth, has improved small-scale irrigation, resulting in higher yields, less soil erosion and deforestation and an increase in socioeconomic status for farmers.

Sources: Interviews with Ministry of Agriculture and Development Bank of Jamaica; Planning Institute of Jamaica (2007), UNDP Jamaica (2012).

1.6 Concluding Recommendations for Decision-Makers: Integrating the Triple Dividend of Resilience in DRM Appraisals

Realising the Triple Dividend of Resilience involves a strategic shift, offering a different perspective on how investments can support policies and objectives beyond DRM. The approach offers an enhanced understanding of the broader economic, social and environmental implications of investing in DRM activities. While loss data, risk models and appraisal tools are the key means for investment decision-making, the overarching foundation of the Triple Dividend of Resilience concept is a more holistic strategy that links DRM, climate and other development policy objectives. Thus DRM is not seen as an objective in its own right—it is considered an important lever for overall development progress that reduces avoided losses and yields benefits from taking risks.

This approach starts with thinking through development strategies and the inherent dynamics of economic development. It then requires the stress-testing of these strategies, based on a range of possible climate futures and the principles of avoiding locking in development paths that are, or may become, unsustainable under climate change.

In practical terms, when making development and DRM plans, policy-makers should resist the temptation (and analytical convenience) of relying on a single set of parameters for analysing risks, costs and benefits. The characteristics of risk are often context-specific and the requirements for assessment differ between local or national scales. Similarly, for hazards with a high probability of recurrence, the measurement of benefit and cost calculations may prove less problematic than for hazards with low and uncertain probabilities (such as earthquakes). We therefore suggest applying multiple approaches and not relying on a single assessment. By way of a conclusion, the following steps indicate some relevant guidance for decision- makers to move towards the Triple Dividend of Resilience perspective (Garrido 2015).

1.6.1 Define the Problem and Its Context

A practical starting point for decision-makers is a mapping exercise to understand development goals, threats and risk drivers:

- What are the contextual development goals set by a certain country, city, locality or village?
- What are the threats to, and drivers of, development?
- What DRM measures are proposed and how do they relate to these goals, threats and drivers?
- Who are main beneficiaries? To what extent are individuals, groups, sectors or activities better protected because of DRM?

1.6.2 *Identify and Apply Tools and Methods for Empirical Analysis of DRM*

Ideally, a DRM proponent should select a set of approaches that can generate quantitative measures or shed light on each of the three types of dividends of resilience. It is unlikely that a single approach can yield answers to every single benefit stream linked to DRM. A more complete evaluation requires the use of various qualitative and quantitative assessment tools. The application of multiple approaches rather than reliance on one tool or method is recommended, especially in data-constrained environments, where flexible approaches are needed:

- Conduct a probabilistic assessment rather than rely only on historic loss figures. This can yield clearer understanding of the first dividend (saving lives and avoiding losses).
- Using simple proxies to measure the second dividend of resilience may be necessary. The biggest gap in triple dividend knowledge lies in understanding how reducing background risk can help unlock and stimulate economic activity. Anticipated land value increase could be used as a good estimate of increased economic activity in a given project area, for example. Another more sophisticated option would be to identify risk thresholds and acceptable levels of risks for different stakeholders.
- The economic value of dual-purpose infrastructure, as well as possible cost savings, can be used to measure the value of the third "co-benefits" dividend. Assessments to monetise non-market values may also be required to widen the scope of assessments of social and environmental co-benefits.

1.6.3 *Communicate Outcomes*

Communicating the triple dividend assessments to other stakeholders, such as business, tax payers and political supporters, is an essential requirement for integrating the concept into development planning:

- Communicate how DRM interventions are linked to, or can be delivered through, other development policies and interventions. Explain the benefits of DRM actions using triple dividend principles and the value of DRM interventions relative to "do nothing" scenarios.
- Focus on supporting development paths that are robust to a range of possible climate and socioeconomic futures. Recognising the need to integrate DRM into future development pathways, to curb the rise of disaster losses, constitutes an important step towards achieving sustainable development objectives.

- Devise strategies for communicating the dividend concept. This includes communicating how DRM interventions are linked, or can be delivered through, other development interventions. What are the benefits from DRM under triple dividend principles and are they robust under different climate and development futures? What is the value of DRM interventions relative to "do nothing" scenarios?
- Identify the implications of fear and risk aversion. The experience of a disaster and the ever-present background risk of future disasters can hamper development and cause economic paralysis. The biggest gap in triple dividend knowledge is in understanding how mitigating such background risk can help unlock and stimulate economic activity. While quantification of these effects is highly case-specific, one option would be to identify risk thresholds and acceptable levels of risks for different stakeholders.

References

Auker M, Sparks R, Siebert L, Crosweller H, Ewert J (2013) A statistical analysis of the global historical volcanic fatalities record. J Appl Volcanol 2(2). doi:10.1186/2191-5040-2-2

Carter M, Barrett C (2006) The economics of poverty traps and persistent poverty: an asset-based approach. J Dev Stud 42(2):178–199

CDP (Carbon Disclosure Project) (2015) Climate change resilience in europe: a snapshot of the private sector. CDP, London

Crawford M, Seidel S (2013) Weathering the storm: building business resilience to climate change. Center for Climate and Energy Solutions, Arlington, VA

Dar M, de Janvry A, Emerick K, Raitzer D, Sadoulet E (2013) A flood of investments: flood-tolerant rice and farm investment in India. University of California, Berkley, CA

Dercon S (ed) (2005) Insurance against Poverty. Open University Press, Oxford

Díaz-Perera MÁ (2013) La construcción de las condiciones históricas de posibilidad de un desastre: el caso de dos colonias de Villahermosa, Tabasco. Final report of FORDECYT Cuenca Grijalva Project. El Colegio de la Frontera Sur, Villahermosa, Mexico

Doswald N, Munroe R, Roe D, Giuliani A, Castelli I, Moller I, Stephens J, Spencer T, Vira B, Reid H (2014) Effectiveness of ecosystem-based approaches for adaptation: review of the evidence-base. Clim Dev 6(2):185–201

Elbers C, Gunning JW, Kinsey Bill (2007) Growth and risk: methodology and micro evidence. World Bank Econ Rev 21(1):1–20

Emerick K, de Janvry A, Sadoulet E, Dar M (2015) Technological innovations, downside risk, and the modernization of agriculture. Poverty Action Lab, Cambridge, MA

ERM (Environmental Resources Management) and DFID (Department for International Development) (2005) Natural disaster and disaster risk reduction measures. A desk review of costs and benefits. ERM, London

Field C, Barros V, Stocker T, Dahe Q, Dokken DJ, Ebi KL, Mastrandrea MD, Mach KJ, Plattner G-K, Allen SK, Tignor M, Midgley PM (eds) (2012) Summary for policy makers. In: Special report on managing the risks of extreme events and disasters to advance climate change

adaptation: summary for policymakers. Report of Working Groups I and II of the IPCC. Cambridge University Press, Cambridge and New York
Fleisher B, Chen J (1997) The coast-noncoast income gap, productivity, and regional economic policy in China. J Comp Econ 25(2):220–236
Gallup J, Sachs J, Mellinger A (1998) Geography and economic development. Working Paper 6849. NBER, Cambridge, MA
Garrido L (2015) Ideas towards the building guidelines for quantitative analysis of DRMR. Unpublished discussion note to GFDRR, World Bank
GFDRR (Global Facility for Disaster Reduction and Recovery) (2014) Building back better in Tonga after Cyclone Ian. Stories of Impact Briefing. World Bank, Washington, DC
GFDRR (Global Facility for Disaster Reduction and Recovery) (2015) Investing in resilience. World Bank, Washington, DC
Ghesquiere F, Mahul L, Jamin O (2006). Earthquake vulnerability reduction program in Colombia: a probabilistic cost-benefit analysis. Policy Research Working Paper. World Bank, Washington, DC
Global Commission on the Economy and Climate (2014) New climate economy technical note: infrastructure investment needs of a low-carbon scenario. Global Commission on the Economy and Climate, Washington, DC
Global Commission on the Economy and Climate (2015) Seizing the global opportunity, the new climate economy report 2015. Global Commission on the Economy and Climate, Washington, DC
Gollier C, Pratt J (1996) Risk vulnerability and the tempering effect of background risk. Econometrica 64(5):1109–1123
Greatrex H, Hansen J, Garvin S, Diro R, Blakeley S, Le Guen M, Rao K, Osgood D (2015). Scaling up index insurance for smallholder farmers: recent evidence and insights. Report 14. CCAFS, Copenhagen
Hallegatte S (2014) Economic resilience: definition and measurement. Policy Research Working Paper 6852. World Bank, Washington, DC
Hoflinger R, Mahul O, Ghesquiere F, Perez S (2012) FONDEN: Mexico's natural disaster fund—a review. World Bank, Washington, DC
IFRC (International Federation of Red Cross and Red Crescent Societies) (2012) Mangrove plantation in Viet Nam: measuring impact and cost benefit. Case Study. IFRC, Geneva
Jones H, Hole D, Zavaleta E (2012) Harnessing nature to help people adapt to climate change. Nat Clim Change 2(7):504–509
Keefer P (2009) Disastrous consequences: the political economy of disaster risk reduction. Working Paper. World Bank, Washington, DC
Kellett J, Caravani A (2013) Financing disaster risk reduction: a 20 year story of international aid. ODI, London
Khan MSA (2008) Disaster preparedness for sustainable development in Bangladesh. Disaster Prev Manag Int J 17(5): 662–671
Kok M, Metz B, Verhagen J, van Rooijen S (2008) Integrating development and climate policies: national and international benefits. Clim Policy 8(2):103–118
Kopits G (2014) Coping with fiscal risk. OECD J Budgeting 14(1):47–71
Lovell E, le Masson V (2014) Equity and inclusion in disaster risk reduction: building resilience for all. CDKN and ODI, London
Lovett R (2010) Why chile fared better than Haiti. Nature Briefing, 10 March. Nature Publishing Group, London
Madajewicz M, Tsegay A, Norton M (2013). Managing risks to agricultural livelihoods: impact evaluation of the HARITA program in Tigray, Ethiopia, 2009–2012. Evaluation Report. Oxfam America, Boston, MA

Mechler R, Bouwer L (2015) Understanding trends and projections of disaster losses and climate change: is vulnerability the missing link? Clim Change 133(1):23–35

Moody's (2015) Flood risk in coastal Virginia supports need for proactive planning, capital investments. Announcement, 18 June. Moody's, New York

Patankar A (2015) The exposure, vulnerability, and ability to respond of poor households to recurrent floods in Mumbai. Policy Research Working Paper 7481. World Bank, Washington, DC

Peterson, G (2012) Unlocking land values to finance urban infrastructure. PPIAF Trends and Policy Options 7. World Bank, Washington, DC

Phaup M, Kirschner C (2010) Budgeting for disasters: focusing on the good times. OECD J Budgeting 2010(1):1–24

Planning Institute of Jamaica (2007) The poverty-environment nexus: establishing an approach for determining special development areas in Jamaica. Sustainable Development and Regional Planning Division, Planning institute of Jamaica, Kingston

Rentschler J (2013) Why resilience matters: the poverty impacts of disasters. Policy Research Working Paper 6699. World Bank, Washington, DC

Rodin J (2014) The resilience dividend: being strong in a world where things go wrong. Public Affairs, Washington, DC

Rogers D, Tsirkunov V (2011) Costs and benefits of early warning systems. Background Paper for Global Assessment Report. UNISDR, Geneva

Santucci L, Puhl I, Sinha A, Enayetullah I, Agyemang-Bonsu W (2015) Valuing the sustainable development co-benefits of climate change mitigation actions. UNESCAP, Bangkok

Scor SE (2013) Supply chain and contingent business interruption (CBI): a perspective on property and casualty. Scor Global P&C Focus, December. SCOR SE, Paris

Seang, H (2009). A case study of mitigating flooding in city center of Kuala Lumpur. In: Innovative strategies towards flood resilient cities in Asia-Pacific. UNESCAP, Bangkok

Shepherd A, Mitchell T, Lewis K, Lenhardt AJ, Lindsey SL, Muir-Wood R (2013) The geography of poverty, disasters and climate extremes in 2030. ODI, London

Standard & Poor's Rating Services (2014) Special report: climate change. Preparing for the long term. Standard & Poor's Ratings Services: Credit Week, The Global Authority on Credit Quality. 28 May 2014. Standard & Poor's Rating Services, McGraw Financial, New York

Standard & Poor's Rating Services (2015) Climate change will likely test the resilience of corporates' creditworthiness for natural catastrophes. Ratings Direct. 20 April 2015. Standard & Poor's Rating Services, McGraw Financial, New York

Subbiah A, Bildan L, Narasimhan R (2008) Assessment of the economics of early warning systems for disaster risk reduction. Background Report submitted to GFDRR. Regional Integrated Multi-Hazard Early Warning System, Asian Disaster Preparedness Center, Bangkok

Surminski Swenja (2014) The role of insurance in reducing direct risk: the case of flood insurance. Int Rev Environ Resour Econ 7(3–4):241–278

Tanner T, Lewis D, Wrathall D, Bronen R, Cradock-Henry N, Huq S, Lawless C, Nawrotzki R, Prasad V, Rahman MA, Alaniz R, King K, McNamara K, Nadiruzzaman M, Henly-Shepard S, Thomalla F (2015) Livelihood resilience in the face of climate change. Nat Clim Change 5:23–26

The Economic Times (2011) Honda to cut production at Indian arm by Half. The Economic Times, 26 April

Tinputz District DRM Committee (2014) Tinputz district disaster risk management plan. Tinputz District DRM Committee, Tinputz

UNDESA (UN Department of Economic and Social Affairs) (2014) Revision: world urbanization prospects. UNDESA, New York
UNDP (UN Development Programme) Jamaica (2012) Water is life: empowering St. Elizabeth farmers. UNDP Stories. UNDP Jamaica, Kingston
UNISDR (UN Office for Disaster Risk Reduction) (2015a) Making development sustainable: the future of disaster risk management. Global assessment report on disaster risk reduction. UNISDR, Geneva
UNISDR (UN Office for Disaster Risk Reduction) (2015b) Chile's investment in disaster risk reduction pays off. News Archive, 17 September. UNISDR, Geneva
Van Tran GP (2013) Lessons from Typhoon Nari. Storm resistant housing shown to be effective. Institute for Social and Environmental Transition-International, Hanoi
WEF (World Economic Forum) (2015) The global risks report 2015. WEF, Davos
Wilkinson E (2012) Transforming disaster risk management: a political economy approach. Working and Discussion Papers. ODI, London
World Bank (2012) Mekong integrated water resources management project. Project appraisal document. World Bank, Washington, DC
World Bank (2013) World development report 2014. Risk and opportunity: managing risk for development. World Bank, Washington, DC
World Bank (2014a) A novel approach to disaster risk management: the story of Mexico. World Bank, Washington, DC
World Bank (2014b) Implementation completion and results report, Vietnam natural disaster risk management project. World Bank, Washington, DC
World Bank (2015) Global monitoring report 2015/2016. World Bank, Washington, DC
WRI (World Resources Institute) (2008) Roots of resilience—growing the wealth of the poor. WRI, Washington, DC

Chapter 2
Avoided Losses and the Development Dividend of Resilience

Stéphane Hallegatte, Mook Bangalore and Marie-Agnès Jouanjean

Abstract Global economic losses from natural disasters continue to increase. And yet investments in disaster risk management (DRM) are not universal, as they are traditionally seen as being in competition with other development and economic priorities, and the multitude of benefits of DRM investments are not traditionally accounted for in cost–benefit analyses. This chapter contributes to this discussion by highlighting the multiple benefits of DRM investments, focusing on both the avoided losses when a disaster occurs but also the impacts on economic development even before a disaster strikes. The main message is that as well as reducing losses when a disaster strikes, DRM investments can generate a second 'development' dividend of resilience through a shift of investment strategies and perhaps even an increase in investment value that could benefit the economy even before a disaster strikes. Providing evidence about the existence of both these dividends to policy-makers and investors can contribute to a narrative reconciling short- and long-term objectives, thereby improving the acceptability and feasibility of DRM investments.

Keywords Disaster risk management · Resilience · Development · Economic losses

2.1 Introduction

Global economic losses from natural disasters are increasing over time and in 2014 totalled $110 billion (Munich 2015). There have been repeated calls to do more to prevent disasters or minimise their consequences for affected populations.

S. Hallegatte (✉) · M. Bangalore
Climate Change Group, World Bank, Washington, DC, USA
e-mail: shallegatte@worldbank.org

M. Bangalore
e-mail: mbangalore@worldbank.org

M.-A. Jouanjean
Organization for Economic Cooperation and Development, Paris, France
e-mail: Marie-Agnes.JOUANJEAN@oecd.org

S. Surminski and T. Tanner (eds.), *Realising the 'Triple Dividend of Resilience'*, Climate Risk Management, Policy and Governance,
DOI 10.1007/978-3-319-40694-7_2

Investments in disaster risk management (DRM), however, have to compete with many other development and economic priorities, and economic analyses of the cost and benefits of disaster risk investments have been the focus of intense research and discussions.

This chapter discusses the benefits of DRM expenditures, particularly those that go beyond avoidance of losses of lives and assets through better risk management. We focus on developing the narrative and providing evidence for the first dividend (reducing economic losses) and on the second dividend (development) of the framework outlined in Chap. 1. Chapter 3 discusses benefits linked to the third dividend (on co-benefits). Our main message is that DRM investments can also reduce indirect losses when a disaster strikes, and can lead to a shift of investment strategies and potentially an increase in investment value that could benefit the economy even before or in the absence of a disasters. However, at present, these "benefits" of DRM are understated and often not included in cost–benefit assessments. Providing evidence on the existence of these dividends to policy-makers and investors can help generate a narrative reconciling short- and long-term objectives, thereby improving the acceptability and feasibility of DRM investments.

2.2 Higher Disaster Losses at the Macro Level

Total economic losses from a disaster can be much larger than the face value of what is affected (asset damages), and the so-called "ripple effect" means its impacts extend beyond the directly affected population and infrastructure. For instance, certain economic sectors or segments of the population not hit by a storm can experience less income because of lower demand. These indirect impacts can affect long-term prospects of economic growth and development both in the geographic area in which the shock occurred but also elsewhere, according to the level and scale of economic integration. Accounting for and including these indirect losses in a DRM cost–benefit analysis is often challenging as a result of a shortage of data. It also requires recognition of certain moderating factors that determine how strong these indirect impacts are (Table 2.2).

2.2.1 Indirect Losses from the Disruption of Economic Infrastructure and Activity

Disasters are "macro" events as they can affect all economic actors in the area where they occur—households, government agencies and firms—even those that do not experience any material or human losses. Smith and McCarty (2009), investigating the impact of the 2004 hurricane season in Florida on household displacement, found that among the 21 % of households forced to move out of their home

Table 2.1 Reason for business closure following the 1994 Northridge earthquake

Reason	% of firms reporting	Local (L) or indirect (I)
Needed to clean up damage	*65.2*	*Local*
Loss of electricity	58.7	Indirect
Employees unable to get to work	56.4	Indirect
Loss of telephones	49.8	Indirect
Damage to owner or manager's home	44.4	Indirect
Few or no customers	39.9	Indirect
Building needed structural assessment	*31.5*	*Local*
Could not deliver products or services	24.0	Indirect
Loss of machinery or office equipment	*23.7*	*Local*
Building needed repair	*23.4*	*Local*
Loss of inventory or stock	*21.9*	*Local*
Loss of water	18.2	Indirect
Could not get supplies or materials	14.9	Indirect
Building declared unsafe	*10.1*	*Local*
Could not afford to pay employees	*9.5*	*Local*
Loss of natural gas	8.7	Indirect
Loss of sewer or waste water	5.3	Indirect
Other	15.8	Both

Note Reasons linked to local damages to the business are highlighted in italics; others are indirect reasons, owing to perturbations in infrastructure services such as transport or electricity
Source Tierney (1997)

after a disaster, 50 % had to do so because of the loss of utilities (e.g. no running water). Only 37 % had to move because of structural damage to their house.

Tierney (1997) found that the loss of utility services and transport following the 1994 Northridge earthquake in Los Angeles, California, had a heavy impact on firms. According to her, 65 % of the small businesses investigated closed after the earthquake because of the need to clean up damages. The five other most important reasons, mentioned by 59–40 % of the sample, included loss of electricity, employees' inability to get to work, loss of telephones, damage to owner's home and reduction in demand, with few or no customers (Table 2.1). Such issues are not related to direct structural damages to the business itself but to off-site impacts.

Business activity does not occur in isolation. Businesses are often integrated in a value chain and depend on upstream and downstream activities and stakeholders. Therefore, owing to complex economic intricacies, business output losses can be the consequence of a shock to the economic activity both upstream (backward) and downstream (forward), and the creation of bottlenecks within supply chains.[1]

[1]These ripple effects can even take place within a factory, if one segment of the production process is impossible and therefore interrupts the entire production.

According to the position of the bottleneck in the value chain, ripple effects can be backward or forward:

- *Backward ripple effects* arise when a shock propagates from clients to suppliers. For example, if the production of a client is incapacitated, input demand to its suppliers will also reduce. For suppliers sales will reduce, despite the absence of direct damages to its production capacity.
- *Forward ripple effects* arise when the impact propagates from suppliers to clients, for example when a client is open for operation but its supplier is unable to produce or sell inputs needed for production processes.

The output losses from a disaster depend on firm-to-firm network characteristics such as average number of suppliers, degree of complementarity and shape and structure of connections between firms (Henriet et al. 2011). Modern organisation of production, characterised by international production networks,[2] a limited number of suppliers, small stocks and production on demand, has created new forms of vulnerabilities to natural disasters, well beyond domestic economy frontiers. The impact of disasters on global value chains was illustrated by the Tohoku-Pacific earthquake in Japan in March 2011, and its consequences for domestic industrial production and the resulting decrease in exports of goods used as inputs, for instance in the auto industry. *The Economic Times*, an Indian newspaper, reported that, "Japan's Toyota Motor will cut production at its Indian subsidiary by up to 70 % between April 25 and June 4 due to disruption of supplies" (The Economic Times 2011).

If an economy's capital stock consists of a bundle of complementary assets, the destruction of one component reduces the overall productivity of the entire production system with an indirect impact much larger than what could be expected from the analysis of one destroyed component only. One relatively straightforward example illustrating the difference between direct and indirect losses is given by the case of two cities connected by a single road. Destruction of only a segment of this road is enough to disrupt freight connections between those two cities. The loss resulting from the destruction of one segment can therefore not be estimated based on the value of this segment, but requires an analysis of the entire production system depending on the connection between the two cities. The same is true—to some degree—for the entire economic system: the loss of one asset will have repercussions for others that depend on it.

Past disasters provide useful examples. The San Francisco–Oakland Bay Bridge, essential to both cities' economic activity, was closed for one month after the 1989 Loma Prieta earthquake (Fig. 2.1). This closure affected almost all small and large business in the Bay Area (Kroll et al. 1991) and, although it was difficult to quantify losses in economic activity, the scale of output losses was an order of magnitude higher than the amount needed to repair the bridge. The health care system in New

[2]We alternatively use the expressions "international production networks" and "global value chains".

Fig. 2.1 The Oakland–San Francisco Bay Bridge, which was closed for one month following the 1989 Loma Prieta earthquake. *Source* Dan Bluestein, Wikimedia Commons

Orleans is another example. Beyond the immediate economic value of the service it provides, a functioning health care system creates positive externalities, acting for example as a pull factor attracting workers to the region. After Katrina's landfall in 2005, the health care system experienced significant disruption and did not recover quickly (Hallegatte 2008; Rudowitz et al. 2006). Poor health care services made it more difficult to attract construction workers to the region (indeed, construction is a high-risk occupation), slowing down the reconstruction process. As a consequence, the disruption of health care services in the hurricane and its aftermaths went beyond the loss of its asset value.

It is important to note that not all indirect impacts are negative. Disasters reduce production capacity, but also increase demand for outputs from the reconstruction sector. Thus, reconstruction can act as a stimulus. However, the resulting dynamic depends on pre-existing economic conditions, such as the phase of the business cycle and the existence of distortions that lead to under-utilisation of production capacities (Hallegatte and Ghil 2008). If the economy is efficient and in a phase of high growth, in which all resources are fully used, the net effect of a stimulus will be negative, for instance through diverted resources, production capacity scarcity and accelerated inflation. If the pre-disaster economy is depressed, however, the stimulus effect may in some cases (e.g. when there are substantial aid flows) yield benefits to the economy by mobilising idle capacities.

For instance, in 1992, when Hurricane Andrew hit south Florida, the region's economy was sluggish, with 50 % unemployment among construction workers (West and Lenze 1994). Reconstruction had a large stimulus effect in the economy, which would have been impossible in a better economic situation such as the one in 2004, when four hurricanes hit Florida during a housing construction boom (ibid.).

Finally, old and low-quality construction is generally more vulnerable to damages than more recent capital. In the case of a disaster, the destruction of low-quality assets may allow the possibility of "building back better", improving the situation post-disaster. For instance, an earthquake may destroy old, low-quality, buildings, making it possible to rebuild with improved building norms. For example, after the Christchurch earthquake in New Zealand in 2011, building norms for energy efficiency led to better comfort and lower energy bills (Miles et al. 2014). However, experiences from the reconstruction process in Haiti after the 2010 earthquake found building back better may be much more difficult in practice, owing to a lack of adequate funding and technical expertise and raw materials in the disaster location (Kijewski-Correa and Taflanidis 2011).

2.2.2 *Impact on Long-Term Growth and Development*

Natural disasters have economic impacts, which extend beyond the short and medium run and affect long-run growth. Reconstruction indirectly affects the economy by crowding out consumption and investment. Post-disaster, uninsured households divert consumption towards reconstruction or draw down savings, potentially reducing the availability of investments in the economy (Hallegatte 2014). The same is true for firms, which have to divert investments and profit redistribution to households towards reconstruction spending. This effect can have a broad, economy-wide depressing impact. Ranger et al. (2011) find that the total indirect effect to the economy from the 2005 floods in Mumbai, India, would have been halved (reduced by $200 million) if all losses had been paid through insurance instead of letting households use their savings and firms their own resources, as occurred.

While such diversion can potentially have a negative effect on the economy, so can a lengthy reconstruction process, which depends on the degree and capacity to divert funds away from investments and consumption. While the €10 billion spent on reconstruction expenditure following the 2002 floods in Germany corresponds only to the equivalent amount of investments spent over 10 days in the country, reconstruction was spread out over more than three years, suggesting only a small fraction of investments can be dedicated towards reconstruction.

Therefore, reconstruction processes might become lengthier than expected, as consumers, insurance and reinsurance companies, firms and public organisations need time to direct large amounts of money to reconstruction, a constraint especially stringent in developing economies that are already lacking financial service infrastructure and lagging behind in investment capacity (Benson and Clay 2004).

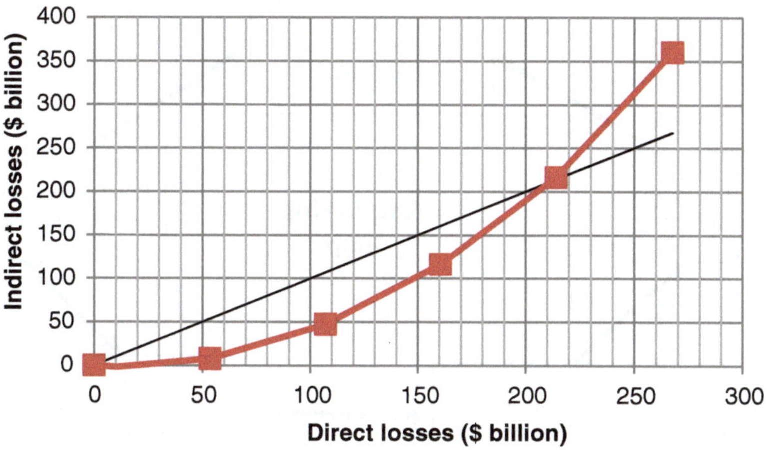

Fig. 2.2 Indirect (output) losses as a function of direct (asset) losses in Louisiana for Katrina-like disasters of increasing magnitude. *Note* The *red curve* signifies indirect losses. *Source* Hallegatte (2008)

Another source of friction is the reconstruction sector's capacity to absorb the increase in demand following a disaster: skill availability and organisational capacities are adapted to the normal state of affairs and are not always able to face huge increases in demand. One illustration of this issue relates to the long reconstruction periods that followed the French storms in 1999 and the AZF factory explosion in Toulouse in 2001, owing to as shortage of roofers and glaziers.

Therefore, the extent of indirect losses owing to the destruction of productive assets and infrastructure in terms of economic activity and growth does not depend only on the physical intensity of the natural event but also on the coping capacity of the affected human system and its ability to rebuild rapidly and efficiently. While investment spillovers are not an asset "loss", in the absence of tools to better manage risk and reallocate resources post-disaster economic losses are certainly higher.

Hallegatte (2008) models the direct and indirect losses from Katrina-like disasters in Louisiana. A non-linear relationship emerges: when direct losses are less than $50 billion, reconstruction is rapid and aggregated indirect losses stay close to zero.[3] Beyond $50 billion of direct losses, the reconstruction period extends over several years and indirect losses increase exponentially. When direct losses exceed $200 billion, total losses are twice as large as direct losses (Fig. 2.2).

Such non-linear relationships lead to large and long-term reductions in growth and lost output and may lead potentially to macro-level poverty traps, with entire regions falling into a vicious circle, leading the economy toward a lower growth equilibrium and reducing development capacity (Hallegatte et al. 2007). Such

[3]Note the aggregation hides important disparities among sectors and social categories.

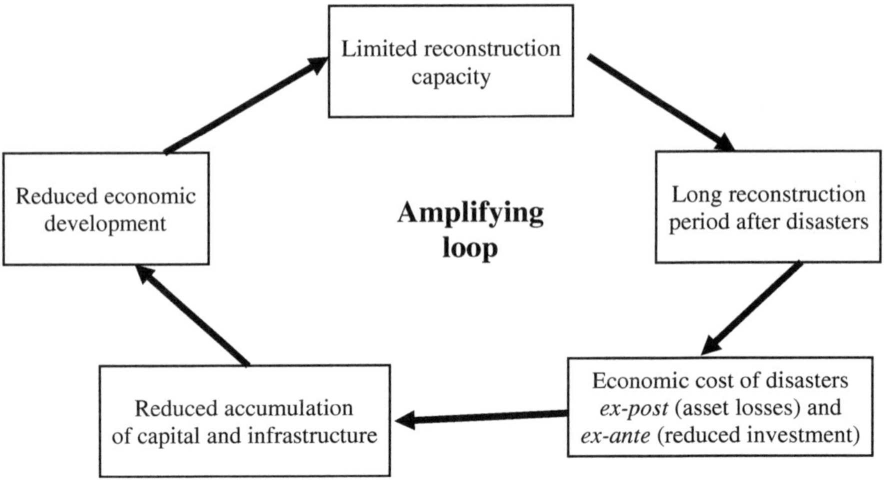

Fig. 2.3 Amplifying feedback loop that illustrates how natural disasters could potentially become responsible for macro-level poverty traps

poverty traps can be explained by amplifying feedbacks, as presented in Fig. 2.3. Many regions have limited capacity to rebuild after a disaster. If the region is regularly affected, it may not have enough time and resources to rebuild its asset base between two events. As a result, it may end up in a permanent state of reconstruction, allocating resources to rebuild rather than investing in new additional infrastructure and equipment, preventing capital accumulation and infrastructure development. Such a cycle, in the absence of external intervention, can lead to permanent disaster-related underdevelopment.

Over the long run, the effect on economic growth is a balance between negative and positive spillovers. The analysis of this shows ambivalent results, suggesting disasters have differential macroeconomic impacts, determined by a variety of factors, such as the absorptive capacity of an economy and its access to international capital including aid due, as well as the scale of a disaster (For contributions to the discourse, see Albala-Bertrand 1993; Felbermayr and Gröschl 2014; Strobl 2010; also Chap. 4).

Through interruptions of infrastructure and baseline services, propagations in the supply chain and diversion of spending by households and firms towards reconstruction, disaster losses go well beyond the direct asset destruction and affect the overall macroeconomics dynamics. These indirect effects, and what moderates them, are summarised in Table 2.2. Furthermore, the most recent literature on the impact of disasters on growth suggests that, while small disasters may not have long-term macroeconomic consequences, large ones are likely to have measurable long-term negative effects on economic growth (Felbermayr and Gröschl 2014; Hsiang and Jina 2014; Loayza et al. 2012).

Table 2.2 Summary of the indirect effects of natural disasters at the macroeconomic scale

Type of indirect effect	Moderated by
Losses in electricity and transport	Infrastructure quality and reliability
Supply chain ripple effects	Complementarity and size of shock
Crowding out investment	Level of insurance penetration
Stimulus	Existing economic situation
Capital replacement	Type of capital replacement

2.3 Welfare Losses at the Microeconomic Level

The previous section that focusing the evaluation of macroeconomic losses from a disaster on direct losses can be misleading and leads to underestimating the welfare impact. But underestimation of the welfare impact can also arise from disregarding the distributional impacts of disasters. For instance, it seems rather intuitive to think that the impacts of disasters on the livelihoods of poor and marginalised people are more substantial, first because of their higher exposure to physical risks but also because of the reliance of their livelihood strategy on fewer and more vulnerable assets. While the impact of the disaster can be disastrous for such people, the repercussions for gross domestic product can be invisible, especially if the very poor own close to nothing.

Thus, to more precisely examine the impacts of a disaster at the micro level, it is important to examine who is affected and how. Below, we examine first how asset losses are distributed among the population and then how asset losses translate into welfare losses.

2.3.1 Asset Losses Differ Depending on Who Is Hit

Here we examine how asset losses at the microeconomic level are determined and distributed. Asset losses are a function of the hazard, exposure and vulnerability. While a hazard is not determined by socioeconomic characteristics, exposure and vulnerability are.

One major determinant of asset losses is poverty status. First, poor people may be more exposed to natural disasters owing to the role of formal and informal land markets: if natural risks are included in land price valuation (or desirability), poor households should be more likely to live in risky areas where land is cheaper (Fay 2005). This explains why slums are typically located in floodplains or in areas at risk of mudslides, and why poor people are approximately 70 % more likely to be exposed to disasters in cities such as Mumbai (see more in Sect. 2.4) (Patankar 2015). But this may not always be the case. Risky locations may attract richer people: coastal cities are often highly exposed to flood risk, but they host households that are generally richer than those from rural and inland regions, because of

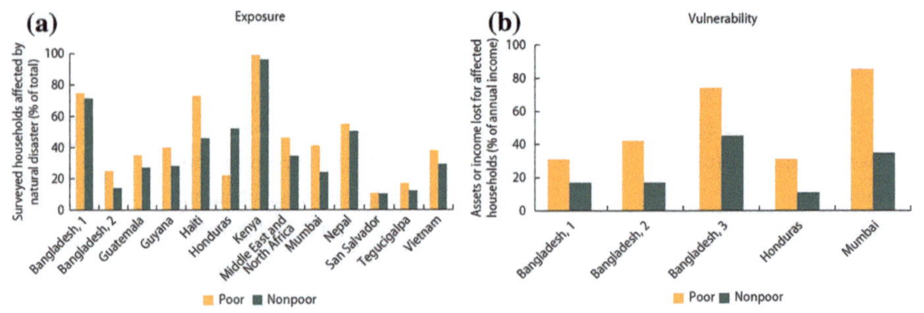

Fig. 2.4 Poverty exposure bias and poverty vulnerability bias exhibited in prior case studies of disaster contexts. *Source* See sources in Hallegatte et al. (2016)

sunny weather and amenities. For instance, Carter et al. (2007) found that Hurricane Mitch in Honduras in 1998 affected only 22 % of households in the poorest quintile, compared with 68 % in the richest quintile. Hallegatte et al. (2016) review case studies of post-disaster contexts examining exposure of poor and non-poor people (see Fig. 2.4 panel a).

While the evidence on poverty exposure to disasters is scale- and context-dependent, it is generally well observed that, when hit, poor people lose more in relative terms. This "vulnerability bias" owes to poor people having lower-quality assets, of which a larger portion are in material form and thus more vulnerable to disasters. For instance, while Carter et al. (2007) found poor people did not have higher exposure to Hurricane Mitch, they were nonetheless more vulnerable in relative terms: poor people lost 31 % of their assets and the rich only 8 %. Hallegatte et al. (2016) review case studies of post-disaster contexts examining the vulnerability of poor and non-poor people (Fig. 2.4 panel b).

The above studies suggest poor people are often more exposed to disasters, and, when hit, lose more. The welfare impacts of disasters can be underestimated by aggregate loss figures, since the value of the assets of poor people are too small to appear in aggregate figures. Therefore, aggregated or averaged asset or output losses do not appear as a metric able to capture the full complexity of disaster outcomes. Instead, welfare losses may be a more appropriate metric. But how to calculate welfare losses from asset losses?

2.3.2 Welfare Losses Are Different from Asset Losses

Taking asset losses as a starting point, two additional areas need to be assessed to estimate welfare impacts: (1) how asset losses translate into income losses and (2) the coping capacity and social protection offered at the individual and government level. Figure 2.5 shows the chain from hazard to welfare impacts.

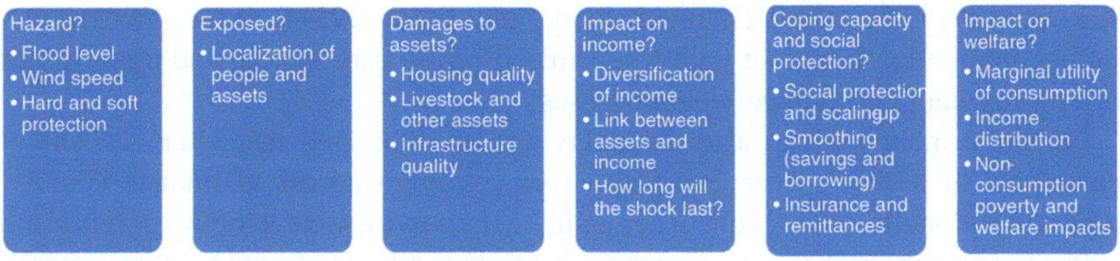

Fig. 2.5 The chain from a natural hazard to its impacts on welfare. *Source* Hallegatte et al. (2015)

The impact of asset damage on income depends on three parameters. The first is related to the reconstruction duration, as described in Sect. 2.2. The second is the link between assets and income (productivity) and the third is the diversification of income.

Estimating how asset losses translate into income losses is difficult to determine. According to the linkages presented in Sect. 2.2, the effect of a disaster on a household's income strategy depends on employment in firms nearby, and, in the case of self-employment, changes in demand for goods. For example, after the 2011 floods in Bangkok, Thailand, Noy and Patel (2014) quantified the direct and spillover effects on income. Households that were directly affected lost on average THB 7600 (approximately $220) in income; households not directly affected by the flood lost almost as much as a result of the reduction in demand or business interruption and ripple effects: THB 6700.

A third component that moderates or magnifies the impact of asset loss on income loss is the diversification of income, including from transfers such as pensions, social protection and remittances. The impact on households' income of the loss of local activities can be smoothed from income sources less affected by a disaster. In particular, government transfers such as pensions and social protection are diversified at the country level, and if a disaster affects only a small part of the country, transfers be only slightly reduced.[4]

Given impacts on income after the disaster, access to private and public coping strategies can reduce the welfare impact of an income loss.

Strategies include private financial mechanisms as well as government targeted interventions and, more generally, social protection. Government transfers can be made available for a period of time after the disaster occurs to allow households to recover from the shock. After the 2005 floods in Mumbai, households received compensation from the government amounting to on average 10 % of household asset losses (Patankar and Patwardhan 2014). However, the compensation scheme did not appear to target poor people or those who lost the most [as was also found

[4]This is true in large countries. However, in small islands, where disasters affect almost all of the population, risk-sharing through diversification may not be an option.

after the 2011 flood in Thailand (Noy and Patel 2014)]. More generally, a socioeconomic environment with a subsidised health care system and opportunities for employment can reduce welfare losses.

Financial inclusion can also help. Savings accounts and insurance can smooth the impact of a shock over time. However, although access to finance is slowly improving globally (van Oudheusden et al. 2015), in most developing countries the ability for poor households to access such services remains limited. More generally in reconstruction contexts, the experience after Cyclone Nargis in Myanmar in 2008 suggests high borrowing rates cripple the speed of recovery (World Bank 2015a).

In an environment with no or little access to social protection and smoothing mechanisms, a shock can potentially lead to poverty traps, especially for asset-poor households. Empirical evidence suggests poor households may liquidate assets in order to cope with shocks and smooth consumption. If the liquidation of assets is insufficient or if shocks are too frequent to rebuild an asset base, households can fall into persistent poverty (Krishna 2006). However, it is also shown that extremely poor households might on the contrary choose to smooth assets rather than consumption (Carter et al. 2007). These households choose to forego consumption rather than further liquidating limited assets in the hope of avoiding poverty traps in the current generation (if their asset base becomes too low, the household may be permanently stuck in poverty). But evidence suggests that such strategies may result in intergenerational poverty traps, as reduced consumption leads to health and educational deficiencies that have impacts on the human capital of children.

Evidence suggests acute impacts on health from lower post-disaster consumption, especially after droughts. Following weather shocks in Sub-Saharan Africa, asset-poor households feed children with less or lower-quality nutrition food, with studies tracking children over decades showing that these behaviours lead to stunting (Alderman et al. 2006; Dercon and Porter 2014; Hoddinott 2006).

Another impact of lower-post disaster consumption on children's human capital occurs through education. Dercon and Porter (2014) found those younger than 36 months at the apex of the famine were less likely to have completed primary school, and estimated the impact to be equivalent to 3 % income losses per year. Intergenerational impacts may endure: recent research in Uganda suggests educated household heads are much less likely to choose coping strategies that involve taking their children out of school (Helgeson et al. 2013).

Investments in DRM before and after a disaster hits can help people manage risks and reduce the welfare impacts of a natural disaster. Better land-use planning and improved building norms can reduce the exposure and vulnerability of the population, and poor people in particular. For instance, land titling in Tanzania is associated with increased housing investments, which reduces vulnerability to floods (Hallegatte et al. 2016). Improved financial inclusion *ex-ante* and favourable financing *ex-post* can hasten recovery and reconstruction. Universal health care insurance and scaled up social protection, for instance through cash transfers, can help avoid detrimental coping strategies taken by households and mitigate shocks. For instance, when droughts in Ethiopia caused food shortages and famine in 2011, the Productive Safety Net Programme expanded its coverage from 6.5 to 9.6 million people in two months and increased the duration of benefits from six to nine months per year (Johnson and Bowen forthcoming).

The take-away from Sects. 2.2 to 2.3 is that, at the macro and micro level, direct asset losses do not tell the whole story. While the impact of a disaster might not be extensive at the macro level, without DRM investments welfare losses can still be substantial fór parts of the population, especially poor people.

2.4 Slower Development in the Absence of DRM Investments

In addition to the loss-centric first resilience dividend, there are further benefits of DRM that arise even in the absence of disaster. This can be in the context of taking "natural" risks when disaster risks are well managed, as well as releasing suppressed economic potential in risky areas.

2.4.1 Development and the Exposure to Natural Hazards

Taking risks is sometimes an unavoidable (or desirable) consequence of development and economic growth. Investing in risky areas can be a conscious and well-informed choice, justified by economic benefits. For instance, increased exposure to natural hazards can be an unavoidable side-effect of investments to create additional employment and growth from international trade in areas characterised by low transportation costs but exposed to flood risks (e.g. Gallup et al. 1998). In China, for instance, Fleisher and Chen (1997) find that Total Factor Productivity (TFP) is 85 % higher in coastal regions than inland, and that TFP growth is not significantly different in spite of higher investment in inland regions, suggesting a permanent productivity advantage in coastal regions from lower transport costs. Cheap waterway transport attracts industrial production close to floodplains, and partly explains why most large cities are located on rivers. In coastal areas, increased exposure to flood can therefore be a deliberate trade-off against higher productivity and economic growth.

The same thing may happen in cities. The drivers of economic growth are concentrated in cities, and productivity growth is larger in cities in part because of positive agglomeration and concentration externality. Ciccone (2002), Lall and Deichmann (2012) and World Bank (2008) report urban–rural income ratios between 1.5 for developed countries and up to 3 for developing countries, suggesting higher productivity in cities at all stages of development. And not only are productivity and consumption higher in urban areas, but also amenities and infrastructure services are often superior: among low-income countries with urban population shares of less than 25 %, access to water and sanitation in towns and cities is around 25 % points higher than it is in rural areas (World Bank 2008). These differences create strong incentives for rapid rural–urban migration. Confronted with land scarcity and high land costs in large cities, this migration has led to construction in at-risk areas (Burby et al. 2001, 2006; Lall and Deichmann 2012). In the most marginal and risky locations, informal settlements and slums are often present, putting poor and vulnerable populations in a situation of extreme risk (Ranger et al. 2011).

An illustrative example of poor people settling in risky areas is Mumbai, which is prone to high flood risk. Patankar (2015) reports on a survey of poor households living in Mumbai's flood areas and shows poor people are well aware of this risk, and are making a deliberate decision to live there to benefit from higher-wage jobs, better schools and medical care and existing social networks. Similar findings are found in Ho Chi Minh City, Vietnam (World Bank and Australian AID 2014).

Risk-taking can also increase welfare through environmental amenities (e.g. from sea views) and generate revenues from tourism. As of 2012, in the Bahamas, Cape Verde, Dominica, Grenada, Macao, Maldives, Montenegro, Samoa, São Tomé & Príncipe, St Lucia, St Vincent and the Grenadines and Vanuatu, tourism accounted for more than half of total exports (World Bank 2015b). Most of these countries are island nations exposed to natural risk (mostly hurricanes and sea level rise), yet expenditures from overseas visitors play a large role in economic output and can hardly be realised without increasing risk.

In situations where there is a trade-off between exposure to natural hazards and productivity or economic growth, improved risk management and more resilient development can mitigate this and accelerate growth and improved productivity (Hallegatte 2014).

This issue relates to the opportunity costs of *ex-ante* risk management, both by households and by lenders. Uninsured risk exposure as well subjective perception endogenously change behaviours, and thus the conditional expected wealth creation dynamics. Failures in financial markets and risk aversion mean the risk of weather-related shocks, including disasters, influences household choices of livelihood strategies in order to minimise the consequences of a shock. Households trade off expected gains for the reduced risk of suffering catastrophic losses.

Such livelihood strategies often entail diversification of activities and less productive investments, constraining productivity and wealth accumulation: households undertake costly behaviours as a means of reducing their exposure to uninsured risk, resulting in forgone welfare gains. Taking into account the prospective consequences of shocks, poor households may manage risk exposure by selecting low-risk, low-return asset and activity portfolios that reduce the risk of greater suffering but limit growth potential and investment incentives (Rosenzweig and Stark 1989). This for instance discourages adoption of new technologies and decreases incentives to invest in productive capital accumulation.

Elbers et al. (2007) provide an illustration of this effect in an agricultural context in Zimbabwe. They found that farmers exposed to risk exhibited a mean capital stock half as large as that for farmers who were not exposed. Of this reduction in capital, *ex-ante* risk accounts for two thirds of the difference. In this case, therefore, most of the welfare impact of risk is through reduced investments and risk-taking, not through damages and losses when the hazard does materialise into an actual event.

2.4.2 Lower Risk-Taking Owing to "Background Risk"

Households and firms face a wide variety of potential shocks that they have to manage together. As an illustration, the 2014 World Development Report (World Bank 2013) reports the frequency of occurrence of a variety shocks, from loss of job to health and floods, in a number of developing countries (Table 2.3). In most countries surveyed, a large proportion of rural households reported being affected by two or more shocks, with drought and flood predominant.

Importantly, the evidence suggests households consider their vulnerability to natural risks like floods and droughts when making other risk-related decisions in other domains—such as creating a business or migrating to a city. Because these risks interact, the existence of natural risk can reduce the willingness to take these other risks, which are necessary for development and growth. Empirical evidence on innovation and entrepreneurship suggests, for instance, that increased risk-taking behaviours are associated with higher economic growth and development:

- The contribution of risk-taking (e.g. through increased innovation/entrepreneurship) to economic growth is well established in the economic literature and was grounded on the theory of endogenous technical change (Romer 1990). The empirical evidence that has followed has largely supported the theory. For innovation, early reviews find a positive link between innovation and output (Cameron 1998; Nadiri 1993). Econometric studies (measuring innovation through patents) provide further support and suggest countries hosting a larger number and higher-quality patents also experience higher economic growth (Hasan and Tucci 2010; LeBel 2008; Yang 2006).
- Regarding entrepreneurship and growth, early studies suggested new business formation promotes employment growth (Birch 1987; Wennekers and Thurik 1999), increased incomes (Carree and Thurik 2002; Picot et al. 1998) and led to greater TFP growth (Aghion et al. 2004; Baumol 2014). In a review of 57 studies, van Praag and Versloot (2007) found entrepreneurial firms had higher productivity growth and increased innovation.
- Furthermore, risk *aversion* has been linked to lower investment in physical and human capital (Rosenzweig and Stark 1989), wage growth (Shaw 1996), and technology adoption (Liu 2012), thereby reducing growth and economic development potential. If high natural risks lead individuals to become less risk-taking in terms of innovation, education or entrepreneurship, growth and development will suffer.

Table 2.3 Households in developing countries face many shocks (% of respondents reporting type of shock)

Shocks	Afghanistan		India	Laos		Malawi		Peru		Uganda	
	U	R	R	U	R	U	R	U	R	U	R
One or more	16.4	48.9	61.6	34.4	72.1	40.0	66.8	20.7	34.4	29.7	56.2
Two or more	8.7	39.2	23.4	11.9	36.1	12.7	40.4	1.4	1.9	5.6	15.6
Natural disasters (drought, flood)	10.6	42.2	57.3	5.6	36.0	10.4	47.2	2.6	21.5	19.9	52.1
Price shocks	0.2	3.0	–	4.4	4.9	21.1	42.0	–	–	1.7	3.2
Employment shocks	6.4	4.3	–	9.3	3.1	7.7	3.4	6.4	1.5	1.9	0.7
Health shocks (death, illness)	6.9	14.0	30.2	23.2	33.8	10.1	18.0	9.1	8.9	11.8	14.9
Personal and property crime	1.8	6.6	0.9	5.8	1.9	8.5	8.4	3.2	3.1	6.6	8.7
Family and legal disputes	–	–	1.9	0.0	0.9	1.7	4.3	0.7	0.3	–	–

Note U urban, *R* rural
Source World Bank (2013) based on data from household surveys, various years 2005–2011

Gollier's seminal work (Eeckhoudt et al. 1996; Gollier and Pratt 1996; Gollier and Schlee 2006) finds, under fairly general conditions, that a higher level of "background risk" (here flood/drought risks) makes individuals less willing to take risks in other domains (e.g. innovation/entrepreneurship). In other words, being exposed to one risk increases an individual's risk aversion regarding other categories of risk. These results suggest households consider their vulnerability to natural risks like floods and droughts when making other risk-related decisions in other domains—such as creating a business or migrating to a city.

Empirical work finds that higher levels of background risk are associated with increased risk aversion in financial decisions (Guiso and Paiella 2008; Lusk and Coble 2008). More recent literature also finds evidence of risk vulnerability with regard to land reform (Tella et al. 2007), early life financial experiences (Malmendier and Nagel 2011), stock market crises (Guiso et al. 2013), and violent trauma (Callen et al. 2014; Voors et al. 2012).

There are two mechanisms through which an increase in the background risk can lead to high risk aversion and lower investment in growth and development.

- The first is rational: there is a possibility that the two independent risks (one related to disasters, the other to risk-taking in general) will materialise together (Gollier and Pratt 1996). This combined risk—and the non-linearity in the utility function—increases risk aversion because a large income shock changes not just an individual's location on the utility function but also the shape of that function (Cassar et al. 2015).
- The second mechanism is behavioural. A shock such as a flood can lead to an overestimation in an individual's perceived likelihood of future natural shocks occurring (Cameron and Shah 2015). Emotional responses can lead individuals to have greater fear of any negative event, reducing risk-taking (Cassar et al. 2015). Consequences, either real or perceived, from multiple shocks occurring in close proximity or simultaneously can be devastating.

The importance of past events on risk aversion is documented in a number of countries, including Bangladesh, Indonesia, Nicaragua and Peru.

Bangladesh. Bangladesh is particularly at risk of coastal flooding and cyclones. Ahsan (2014) examines risk preferences in three coastal communities in Bagerhat, a district in southwest Bangladesh, which regularly experiences cyclones. Socio-economically, the communities studied are heavily reliant on aquaculture and agriculture and are low in income, with average household annual income from farming reported at $1400. Through experiments, risk preferences were investigated and compared with exposure to cyclone. The author found that, on average, non-cyclone-affected subjects bet more in a risk game than subjects who had been affected by cyclones.

Indonesia. East Java, Indonesia, has a population of 37 million and is particularly prone to natural disasters—with floods and earthquakes posing the largest risks. Cameron and Shah (2015) examine whether recent experience of floods and earthquakes affects the level of background risk and risk-taking within the region.

In October 2008, the authors ran a series of experimental games in a random sample of 1550 individuals across 120 villages and found individuals in villages that had suffered a flood/earthquake in the previous three years exhibited higher levels of risk aversion compared with individuals in villages that had not experienced a disaster (41 % decrease in probability of making a risky choice in the experiment).

A year later, the authors conducted a survey asking households to report the probability (or likelihood) that a flood and/or earthquake would occur in their village the following year. For floods (but not earthquakes), individuals who had experienced an event were significantly more likely to report a higher probability of flood in the following year (43 %) compared with those who had not experienced a flood (12 %). Given the true probability of around 3 %, the findings suggest households with recent flood experience over-weight the probability that a future flood occurs. The same is true with severity: those who had experienced a flood also perceived that future floods would be worse. These findings suggest individuals with recent experience perceive the world to be a riskier place; the authors suggest this causes individuals to take fewer risks. Evidence is further provided that behaviour in experiments is correlated with "real-life" risk-taking such as entrepreneurship.

Nicaragua and Peru. Nicaragua and Peru are two disaster-prone countries in Latin America, at risk of flood, drought and hurricane. In 2007, van den Berg et al. conducted risk experiments on a random sample of 100 individuals across regions within each country (Chinandega in Nicaragua; Ancash, Cajamarca, Piura and Tumbes in Peru). They found past experience of a disaster to have a large and significant effect on risk aversion. Across both samples, comparing individuals who had lost assets with those who had not, those who had lost a home exhibited 30 % higher risk aversion; for those who had lost animals this measure was 50 % and for crops it was 60 % (van den Berg et al. 2009). The authors similarly suggest that such reductions in risk aversion continue in the medium run, two years after a disaster. While the authors do not provide evidence on the mechanism through which risk aversion manifests, one plausible conclusion is the increased perception of background risk.

In Vietnam, Reynaud and Nguyen (2012) found experience of floods to have a significant positive effect on demand for insurance, which may reflect higher levels of risk aversion. Also in Vietnam, Dang (2012), combining historical and contemporary survey data, found that individuals living in villages that frequently experience disaster and those who had recently experienced a shock showed higher levels of risk aversion. Abreha (2007) found similar results of drought experience and risk aversion among farmers in Ethiopia.

However, some studies find the opposite—that exposure to natural disasters can make people more risk-loving. In Louisiana, Eckel et al. (2009) conducted an experimental test on individuals exposed to Hurricane Katrina in September 2005, a month after the storm. They found evacuees to be more risk-loving (less risk-averse) after the storm, although this effect was not observed 10 months later. Various other analyses provide evidence on such a change in perception: Page et al. (2014) present a similar risk-loving effect for households that suffered loss as a result of the 2011 Australian floods in Brisbane[5]; Hanaoka et al. (2015) exhibit similar findings using panel data from Japanese households after the 2011 earthquake and tsunami, but only for men, who gamble and drink more after the event[6]; Andrabi and Das (2010) found the 2005 earthquake in Pakistan increased risk aversion; and Said et al. (2014) had a similar result for the 2010 flood in Pakistan.

But not all studies confirm the finding that risk preferences change. Bchir and Willinger (2013), in a field experiment of lahar risk in Peru, found no significant difference of risk aversion between exposed and non-exposed households. Becchetti et al. (2012) had a similar result of no significant difference in a sample of 380 Sri Lankan microfinance borrowers. In addition, preliminary findings from an experimental game in Cambodia's Battambang province actual found experience with a natural disaster had a positive and significant impact on the risk behaviour of participants (Fiala 2016). The contradiction cannot be easily explained by different contexts, since studies disagree even in one given location. Cassar et al. (2015), through risk experiments of 334 subjects from Thai villages affected to different degrees by the 2004 event, found individuals hit hardest by the disaster exhibited strong risk aversion four and a half years after the disaster (in 2009). Callen (2011), in an experiment conducted on a sample of 456 wage workers in July 2007, found no evidence that risk preferences changed. One possibility is that the impact of background risk is more complex than a simple increase in risk aversion. For instance, Li et al. (2011) found individuals exposed to earthquake and snowfall risk in China could not simply be described as more risk-seeking, but that individuals gave more weight to low probabilities after the 2008 China earthquake and snowfall event.

2.5 Conclusion and Implications for Policy

Most investments in DRM still rely on cost–benefit analyses that estimate the benefits from a project or action through the value of the asset (and/or human losses) it can prevent. But indirect losses can be as substantial as direct asset losses; and indirect losses can lead to human losses (e.g. through undernourishment and children stunting) that need to be added to direct human losses.

While these costs are difficult to quantify—and perhaps because so—they are typically excluded from cost–benefit analyses. Nonetheless, the benefits of a DRM policy to reduce indirect losses can be large. Some DRM action even reduces only indirect losses—for instance, insurance and social protection cannot do much to

[5]One limitation of this study is that, relative to the household's situation at the time of survey, the risk game presents only gain options.

[6]The authors found evidence that men became more engaged in gambling and drinking if they were more exposed to the Earthquake.

reduce asset losses, but they minimise the welfare impacts of these losses.[7] Welfare losses can also be much higher than asset losses, when considering the distribution of these losses, and especially the impact on the poorest. And the development benefits from better-managed risks—for instance through the ability to take other risks linked to entrepreneurship or innovation—could also be significant. Yet this is also difficult to quantify and include in a cost–benefit analysis. For example, the benefits people gain from settling in risky areas in urban areas are typically not valued.

Put simply, at present, the "benefits" of DRM are understated—in terms of both avoided losses and increased development. Considering these benefits in policy design is critical to better manage risk.

For example, some actions to reduce risk (or prevent risk generation) may be counterproductive. What is really needed is not *risk reduction*—that would try to reduce the amount of risk-taking indiscriminately—but *risk management*—that prevents excessive risk-taking while allowing risk-taking in cases where the benefits (e.g. proximity to job opportunity) are clear.

For instance, policies that prevent all investments in flood zones in developing countries cities may be extremely costly. They would reduce migration to cities, thereby potentially preventing individuals from accessing higher-pay jobs and better services and children's access to education. It would be more efficient to implement more detailed zoning policies that distinguish between different types of investments to allow worthy ones but prevent inappropriate ones (e.g. by making a difference between housing and production units).

Another option is to invest in safe places. Indeed, it can be rational to experience growing disaster losses only if investments in risky locations are "more" productive than investment in safe places. If investments in transports can make it as desirable to invest in safe places, risk could be reduced without reducing economic growth and output. People in at-risk informal settlements in developing country cities settle there because they face a difficult trade-off between living in risky places with good access to jobs and services and living in a safe place without these opportunities. They would settle in a safe place and reduce flood exposure if better transportation infrastructure and options connected safe living areas to urban opportunities. Similarly, manufacturing plants are created in at-risk coastal areas, but they could be installed in safe areas if transport infrastructure made it possible to ship their production at similar costs. In the broad framework this chapter proposes, transportation investments are risk mitigation investments when they connect safe areas to the opportunities and amenities that currently exist in risky areas.

By providing a strong and holistic risk management framework, a country, a region or even a city makes it possible for all actors to take the risks that are desirable, avoiding excessive risk-taking without constraining growth and development. It also makes it possible to deal with the rare but unavoidable cases where a

[7]Note that well-designed insurance schemes can also create a positive incentive to invest in risk mitigation and prevention.

physical hazard is so violent that it exceeds protection capabilities and causes large losses. In other words, the same DRM policy that reduces welfare losses from a disaster can also provide benefits even before a disaster strikes.

References

Abreha NH (2007) An economic analysis of farmers' risk attitudes and farm households' responses to rainfall risk in Tigray Northern Ethiopia. Ph.D. Thesis, Wageningen University

Aghion P, Blundell R, Griffith R, Howitt P, Prantl S (2004) Entry and productivity growth: evidence from microlevel panel data. J Eur Econ Assoc 2:265–276

Ahsan D (2014) Does natural disaster influence people's risk preference and trust? An experiment from cyclone prone coast of Bangladesh. Int J Disaster Risk Reduction 9:48–57

Albala-Bertrand JM (1993) Political economy of large natural disasters: with special reference to developing countries. Clarendon Press, Oxford and New York

Alderman H, Hoddinott J, Kinsey B (2006) Long term consequences of early childhood malnutrition. Oxford Econ Pap 58:450–474

Andrabi T, Das J (2010) In aid we trust: hearts and minds and the Pakistan earthquake of 2005. World Bank, Washington, DC

Baumol WJ (2014) Stimulating growth amid recession: entrepreneurship, innovation, and the Keynesian revolution. J Policy Model Rapid Growth Stagnation US World Econ? 36:629–635

Bchir M, Willinger M (2013). Does the exposure to natural hazards affect risk and time preferences? Some insights from a field experiment in Perú. Working Paper 13-04. LAMETA, University of Montpellier, Montpellier

Becchetti L, Castriota S, Conzo P (2012) Calamity, aid and indirect reciprocity: the long run impact of tsunami on altruism. Working Paper 316. CSEF, University of Naples, Naples

Benson C, Clay E (2004) Understanding the economic and financial impacts of natural disasters, disaster risk management. World Bank, Washington, DC

Birch D (1987) Job creation in America: how our smallest companies put the most people to work. Scholarly Paper 1496185. SSRN, Rochester, NY

Burby R, Nelson A, Parker D, Handmer J (2001) Urban containment policy and exposure to natural hazards: is there a connection? J Environ Plann Manag 44:475–490

Burby R, Nelson A, Sanchez T (2006) The problems of containment and the promise of planning. In: Birch E, Wachter S (eds) Rebuilding urban places after disaster: lessons from Hurricane Katrina. University of Pennsylvania Press, Harrisberg, PA

Callen M (2011) Catastrophes and time preference: evidence from the Indian Ocean earthquake. Unpublished manuscript. University of California, San Diego, CA

Callen MM, Isaqzadeh JL, Sprenger C (2014) Violence and risk preference: experimental evidence from Afghanistan. Am Econ Rev 104:123–148

Cameron G (1998) Innovation and growth: a survey of the empirical evidence. Nuffield College, Oxford

Cameron L, Shah M (2015) Risk-taking behavior in the wake of natural disasters. J Hum Resour 50:484–515

Carter M, Little P, Mogues T, Negatu W (2007) Poverty traps and natural disasters in Ethiopia and Honduras. World Dev 35:835–856

Carree M, Thurik R (2002) The impact of entrepreneurship on economic growth. Springer, Bern

Cassar A, Healy A, von Kessler C (2015) Trust, risk and time preferences after a natural disaster: experimental evidence from Thailand. Working Paper. University of San Francisco, San Francisco, CA

Ciccone A (2002) Agglomeration effects in Europe. Eur Econ Rev 46:213–227

Dang DA (2012) On the sources of risk preferences in rural Vietnam. Working Paper in Economics and Econometrics 2012-593. College of Business and Economics, School of Economics, ANU, Canberra

Dercon S, Porter C (2014) Live aid revisited: long-term impacts of the 1984 Ethiopian famine on children. J Eur Econ Assoc 12:927–948

Eckel C, El-Gamal M, Wilson R (2009) Risk loving after the storm: A Bayesian-Network study of Hurricane Katrina evacuees. J Econ Behav Organ Individ Decision-Making Bayesian Estimation Market Des A Festschrift in Honor of David Grether 69:110–124

Eeckhoudt L, Gollier C, Schlesinger H (1996) Changes in background risk and risk taking behavior. Econometrica 64:683–689

Elbers C, Gunning JW, Kinsey B (2007) Growth and risk: methodology and micro evidence. World Bank Econ Rev 21:1–20

Fay M (2005) The urban poor in Latin America, directions in development—general. World Bank, Washington, DC

Felbermayr G, Gröschl J (2014) Naturally negative: the growth effects of natural disasters. J Dev Econ Special Issue Imbalances Econ Dev 111:92–106

Fiala O (2016) Experiencing natural disasters: how this influences risk aversion, trust and the demand for microinsurance. Presentation, Dresden, 12 January

Fleisher B, Chen J (1997) The coast-noncoast income gap, productivity, and regional economic policy in China. J Comp Econ 25:220–236

Gallup J, Sachs J, Mellinger A (1998) Geography and economic development. Working Paper 6849. NBER, Cambridge, MA

Gollier C, Pratt J (1996) Risk vulnerability and the tempering effect of background risk. Econometrica 64:1109–1123

Gollier C, Schlee E (2006) Increased risk-bearing with background risk. Working Paper. Department of Economics, W.P. Carey School of Business, Arizona State University, Phoenix, AZ

Guiso L, Paiella M (2008) Risk aversion, wealth, and background risk. J Eur Econ Assoc 6:1109–1150

Guiso L, Sapienza P, Zingales L (2013) Time varying risk aversion. Working Paper 19284. NBER, Cambridge, MA

Hallegatte S (2008) An adaptive regional input-output model and its application to the assessment of the economic cost of Katrina. Risk Anal 28:779–799

Hallegatte S (2014) Natural disasters and climate change. Springer, Cham

Hallegatte S, Ghil M (2008) Natural disasters impacting a macroeconomic model with endogenous dynamics. Ecol Econ 68:582–592

Hallegatte S, Hourcade J-C, Dumas P (2007) Why economic dynamics matter in assessing climate change damages: illustration on extreme events. Ecol Econ 62:330–340

Hallegatte S, Bangalore M, Bonzanigo L, Fay M, Kane T, Narloch U, Rozenberg J, Treguer D, Vogt-Schilb A (2016) Shock waves: managing the impacts of climate change on poverty. World Bank, Washington, DC

Hallegatte S, Bangalore M, Vogt-Schilb A (2015) Socio-economic resilience to floods in 90 countries. World Bank, Washington, DC

Hanaoka C, Shigeoka H, Watanabe Y (2015) Do risk preferences change? Evidence from panel data before and after the great east Japan Earthquake. Working Paper 21400. NBER, Cambridge, MA

Hasan I, Tucci C (2010) The innovation-economic growth nexus: global evidence. Res Policy 39:1264–1276

Helgeson J, Dietz S, Hochrainer-Stigler S (2013) Vulnerability to weather disasters: the choice of coping strategies in rural Uganda. Ecol Soc 18:2

Henriet F, Hallegatte S, Tabourier L (2011) Firm-network characteristics and economic robustness to natural disasters. J Econ Dyn Control 31, 6:150–167

Hoddinott J (2006) Shocks and their consequences across and within households in rural Zimbabwe. J Dev Stud 42:301–321

Hsiang S, Jina A (2014) The causal effect of environmental catastrophe on long-run economic growth: evidence from 6700 cyclones. Working Paper 20352. NBER, Cambridge, MA

Johnson K, Bowen T (Forthcoming) Ethiopia's food security program: social protection, disaster risk management and climate change adaptation

Kijewski-Correa T, Taflanidis A (2011) The Haitian housing dilemma: can sustainability and hazard-resilience be achieved? Bull Earthq Eng 10:765–771

Krishna A (2006) Pathways out of and into poverty in 36 villages of Andhra Pradesh, India. World Dev 34:271–288 (Part Special Issue Corruption and Development: Analysis and Measurement)

Kroll C, Landis J, Shen Q, Stryker S (1991) Economic impacts of the Loma Prieta Earthquake: a focus on small businesses. University of California Transportation Center, Berkeley, CA

Lall S, Deichmann U (2012) Density and disasters: economics of urban hazard risk. World Bank Res Observer 27:74–105

LeBel P (2008) The role of creative innovation in economic growth: some international comparisons. J Asian Econ 19:334–347

Li J-Z, Li S, Wang W-Z, Rao L-L, Liu H (2011) Are people always more risk averse after disasters? Surveys after a heavy snow-hit and a major earthquake in China in 2008. Appl Cogn Psychol 25:104–111

Liu E (2012) Time to change what to sow: risk preferences and technology adoption decisions of cotton farmers in China. Rev Econ Stat 95:1386–1403

Loayza NV, Olaberria E, Rigolini J, Christiaensen L (2012) Natural disasters and growth: going beyond the averages. World Dev 40(7):1317–1336

Lusk J, Coble K (2008) Risk aversion in the presence of background risk: evidence from an economic experiment. In: Cox J, Harrison G (eds) Risk aversion in experiments, research in experimental economics. Emerald Group Publishing, Bingley

Malmendier U, Nagel S (2011) Depression babies: do macroeconomic experiences affect risk taking? Q J Econ 126:373–416

Miles S, Brechwald D, Davidson R, Demeter K, Johnston D, Pampanin S, Wilkinson S (2014) Building back better—case study of the 2010–2011 Canterbury, New Zealand earthquake sequence. Earthquake Engineering Research Institute, Oakland, CA

Munich RE (2015) Review of natural catastrophes in 2014: lower losses from weather extremes and earthquakes. Munich RE, Munich

Nadiri MI (1993) Innovations and technological spillovers. Working Paper 4423. NBER, Cambridge, MA

Noy I, Patel P (2014) Floods and spillovers: households after the 2011 great flood in Thailand. Working Paper 3609. Victoria University of Wellington, School of Economics and Finance, Wellington

Page L, Savage D, Torgler B (2014) Variation in risk seeking behaviour following large losses: a natural experiment. Eur Econ Rev 71:121–131

Patankar A (2015) The exposure, vulnerability and adaptive capacity of households to floods in Mumbai. World Bank Policy Research Working Paper No. 7481, background paper prepared for the report Shock Waves: Managing the Impacts of Climate Change on Poverty, World Bank, Washington, DC

Patankar A, Patwardhan A (2014) Estimating the uninsured losses due to extreme weather events and implications for informal sector vulnerability: a case study of Mumbai, India. World Bank, Washington, DC

Picot G, Manser M, Zhengxi L (1998) The role of self-employment in job creation in Canada and the United States. OECD, Paris

Ranger N, Hallegatte S, Bhattacharya S, Bachu M, Satya Priya K, Dhore FR, Mathur P, Naville N, Henriet F, Herweijer C, Pohit S, Corfee-Morlot J (2011) An assessment of the potential impact of climate change on flood risk in Mumbai. Clim Change 104:139–167

Reynaud A, Nguyen MH (2012) Monetary valuation of flood insurance in Vietnam. Mimeo

Romer P (1990) Endogenous technological change. J Polit Econ 98:S71–S102

Rosenzweig M, Stark O (1989) Consumption smoothing, migration, and marriage: evidence from rural India. J Polit Econ 97:905

Rudowitz R, Rowland D, Shartzer A (2006) Health Care in New Orleans before and after Hurricane Katrina. Health Aff 25:w393–w406

Said, Farah, Uzma Afzal and Ginger Turner. 2014. *Attitudes towards Risk in the Wake of a Rare Event: Evidence from Pakistan.* Working Paper 2-2014. Lahore: CREB, Lahore School of Economics

Shaw K (1996) An empirical analysis of risk aversion and income growth. J Labor Econ 14:626–653

Smith S, McCarty C (2009) fleeing the storm(s): an examination of evacuation behavior during Florida's 2004 Hurricane season. Demography 46:127–145

Strobl E (2010) The economic growth impact of hurricanes: evidence from U.S. coastal counties. Rev Econ Stat 93:575–589

Tella R, Galiant S, Schargrodsky E (2007) The formation of beliefs: evidence from the allocation of land titles to squatters. Q J Econ 122:209–241

The Economic Times (2011) Honda to cut production at Indian Arm by Half. The Economic Times, 26 April

Tierney K (1997) Business impacts of the Northridge earthquake. J Contingencies Crisis Manag 5:87–97

van den Berg M, Fort R, Burger K (2009) Natural hazards and risk aversion: experimental evidence from Latin America. In: International association of agricultural economists conference, Beijing, 16–22 August

van Oudheusden P, Klapper L, Demirguc-Kunt A, Singer D (2015) The global findex database 2014: measuring financial inclusion around the world. World Bank, Washington, DC

van Praag M, Versloot P (2007) What is the value of entrepreneurship? a review of recent research. Small Bus Econ 29:351–382

Voors M, Nillesen E, Verwimp P, Bulte E, Lensink R, van Soest D (2012) Violent conflict and behavior: a field experiment in Burundi. Am Econ Rev 102:941–964

Wennekers S, Thurik R (1999) Linking entrepreneurship and economic growth. Small Bus Econ 13:27–56

West C, Lenze D (1994) Modeling the regional impact of natural disaster and recovery: a general framework and an application to Hurricane Andrew. Int Reg Sci Rev 17:121–150

World Bank (2008) Reshaping economic geography. World development report. World Bank, Washington, DC

World Bank (2013) Risk and opportunity: managing risk for development. World Development Report. World Bank, Washington, DC

World Bank (2015a) Another Nargis Strikes everyday: Post-Nargis social impacts monitoring five years on. World Bank, Washington, DC

World Bank (2015b) World development indicators, international tourism, receipts. World Bank, Washington, DC

World Bank and Australian AID (2014) Where are we during flooding? A qualitative assessment of poverty and social impacts of flooding in selected neighborhoods of HCMC. World Bank and Australian AID, Hanoi

Yang C-H (2006) Is innovation the story of Taiwan's economic growth? J Asian Econ 17:867–8787

Chapter 3
Co-benefits of Disaster Risk Management: The Third Dividend of Resilience

Francis Vorhies and Emily Wilkinson

Abstract Building resilience to climate extremes and disasters can achieve multiple objectives, which can be secondary to the main aim of disaster risk management (DRM) of avoiding disaster losses but can enhance the attractiveness of DRM investments. Co-benefits are often economic, such as investment in dams or irrigation to reduce drought risk, generating greater productivity, but can also be environmental and social. This chapter identifies some of the potential categories of these co-benefits, expanding on typologies created by agencies promoting social and environmental safeguarding in their work. We also look back at previous studies of DRM that mention co-benefits but do not explore them in any detail and examine two new case studies of environmental and socioeconomic co-benefits, one in Jamaica and one in Mexico. We point to a number of challenges in traditional cost–benefit analysis techniques and put forward alternative approaches to identifying environmental and socioeconomic co-benefits when planning DRM investments. We argue that a comprehensive DRM co-benefits framework is needed that includes and categorises all potential positive environmental and socioeconomic impacts. Co-benefits research focused on revisiting existing cases and developing new case studies could play an important role in this regard.

Keywords Disasters · Environment management · Resilience · Sustainable development

---

The authors would like to thank Fernando Aragón and Maurice Mason for providing inputs to the Mexico and Jamaica case studies, and Mook Bangalore and Tom McDermott for their useful comments and suggestions on improving this chapter.

F. Vorhies (✉)
Earthmind, Divonne-les-Bains, France
e-mail: vorhies@earthmind.fr

E. Wilkinson
Overseas Development Institute, London, UK
e-mail: e.wilkinson@odi.org.uk

S. Surminski and T. Tanner (eds.), *Realising the 'Triple Dividend of Resilience'*, Climate Risk Management, Policy and Governance,
DOI 10.1007/978-3-319-40694-7_3

3.1 Introduction

Disasters can be extremely costly, and investing in disaster risk management (DRM) can save both money and lives. Yet policy-makers tend to underinvest or not invest at all in projects to manage risk. This is in part because the costs of such investments are visible and immediate whereas the direct benefits of such investments, and indeed the distribution of these benefits, are unclear and distant.

One way to address this public choice dilemma is to identify, enhance and communicate the co-benefits of DRM. Many *ex-ante* DRM investments will deliver these. They are the benefits that accrue in addition to the primary DRM objectives of avoiding losses and boosting development, and can occur even in the absence of a disaster. Co-benefits can include economic, social and environmental aspects, and be non-DRM-specific (see Chap. 1). The primary objective of DRM investments, or the first resilience dividend, is to avoid disaster losses, but these investments can achieve multiple objectives (Tanner et al. 2015). The co-benefits of investing in DRM, known as the third resilience dividend, are different from the second dividend (unlocking development potential by reducing background risk) (see Chap. 2) in that they are related to specific DRM investments. For example, investments in mangrove reforestation to protect coastal settlements from storm surges can also provide ecosystem services to nearby populations; shelters built for refuge during a storm can also be used as an educational facility. Identifying all these benefits of DRM strengthens the case for DRM investment, but the environmental and social co-benefits are usually less visible and rarely considered when appraising DRM investment.

The notion of co-benefits is relevant beyond disaster and climate risks. Today, investment decisions—both public and private—are increasingly expected to consider their broader contribution to sustainable development within the context of their specific investment focus. This expectation is reflected in the broadening commitment to what is often called 'socially responsible investment'. For example, the institutional investors who have signed up to the Principles for Responsible Investment[1] hosted by the UN Environment Programme (UNEP) Finance Initiative and the UN Global Compact, have committed to "incorporate ESG [environmental, social and corporate governance] issues into investment analysis and decision-making processes".

With respect to DRM, the Rio+20 Outcome Document reaffirms worldwide political commitment to integrating DRM into public and private investments and planning for growth and development:

> We stress the importance of stronger interlinkages among disaster risk reduction, recovery and long-term development planning, and call for more coordinated and comprehensive strategies that integrate disaster risk reduction and climate change adaptation considerations into public and private investment, decision-making and the planning of humanitarian and development actions, in order to reduce risk, increase resilience and provide a smoother transition between relief, recovery and development.

[1]http://www.unpri.org/.

This commitment is further articulated in the integration of DRM into the Sustainable Development Goals (SDGs).[2] Goal 9 is to "build resilient infrastructure, promote inclusive and sustainable industrialization and foster innovation" and includes facilitating "sustainable and resilient infrastructure development in developing countries". This includes a commitment to "upgrade infrastructure and retrofit industries to make them sustainable".

Similarly, proposed Goal 11 aims to "make cities and human settlements inclusive, safe, resilient and sustainable" and brings together environmental priorities such as "efforts to protect and safeguard the world's cultural and natural heritage" and "mitigation and adaptation to climate change", and social priorities such as "access for all to adequate, safe and affordable housing and basic services" and "inclusive and sustainable urbanisation". It also underscores the need for "resilience to disasters" and calls on governments to "develop and implement, in line with the Sendai Framework for Disaster Risk Reduction 2015–2030, holistic disaster risk management at all levels".

The various environmental and social frameworks under development by multilateral institutions—notably the UN and its specialised agencies and the World Bank Group—are also helping establish a set of specific environmental, economic and social issues that can and should be addressed through public and private investments. Investments in DRM should be considered in these frameworks as they generate many of these kinds of co-benefits. They can also generate certain co-costs or unexpected negative externalities, such as those that occur when rivers are dammed upstream to prevent flooding, reducing water for irrigation downstream. Given the opportunities for delivering additional visible returns, ministries of finance and potential public and private co-investors need to ensure such co-benefits are identified and captured to enhance the overall package of returns—direct and indirect—from DRM investments. In so doing, DRM investments will enhance commitment by policy-makers to ensuring growth and development plans and investments are environmental and socially responsible. The successful financing and implementation of DRM programmes and projects in the future is likely to be judged to some extent by their ability to generate environmental and socioeconomic co-benefits, rather than just in terms of losses avoided and lives saved.

3.2 Examples of Local Environmental and Socioeconomic Co-benefits

As each DRM programme or project will face its own unique array of environmental and socioeconomic issues, the opportunities for enhancing co-benefits will also vary. By way of example, one report (DFID 2005) lists a number of examples

[2]http://sustainabledevelopment.un.org/sdgsproposal.html.

of co-benefits that may accrue from different types of DRM investments, divided into four types:

1. Policy and planning, for example institutional, policy and capacity-building measures designed to increase the abilities of countries to manage disaster risks.
2. Physical (prevention), for example building sea walls as part of flood defence mechanisms.
3. Physical (coping/adaptive), for example flood shelters for use during a disaster event.
4. Capacity-building (at the community level), for example developing a disaster preparedness committee.

Table 3.1 outlines some of the potential co-benefits and the co-costs associated with particular DRM investments.

Some of the most commonly cited co-benefits include those associated with community-based DRM. Particularly when efforts to prepare for disasters involve high levels of community participation, these processes can have a positive impact on community cohesion (De Villiers 1999) and even empower vulnerable groups. Other frequently used examples of co-benefits are seen in ecosystems approaches to DRM. The International Union for Conservation of Nature, for example, adopts an ecosystems approach to managing disaster risks that provides benefits to communities in post-disaster situations.[3] Mangrove conservation programmes for DRM protect poor people living on coastal land from storms and also provide wood for fuel. In addition, mangroves are important breeding places for fish and shellfish and contribute to shoreline stability.

Such co-benefits—which materialise irrespective of the advent of a disaster—are clearly important to investment decisions in DRM, although they cannot easily be incorporated in traditional cost–benefit analyses (CBAs). Co-benefits may therefore need to be identified and addressed in their own right, even if they cannot be quantified and included in an aggregate economic analysis of a DRM project.

A review of the literature suggests there may be some co-benefits that are particularly difficult to capture in traditional CBA analysis. These include the following:

- Behaviour change—for example infrastructure designed to reduce risk but with other co-benefit improvements in the design, such as greening and better sanitation
- New business opportunities—for example DRM investments that create new business opportunities, such as a dam that offers opportunities for fishing and more robust fishing boats that permit fishing in deeper waters
- Health benefits and other improvements in well-being—for example DRM investments that improve the urban environment and make people feel better.

[3]http://www.iucn.org/about/work/programmes/ecosystem_management/disaster/about_drr/.

Table 3.1 A review of co-benefits associated with DRM measures

DRM activity	Potential co-benefits	Potential co-costs
Flood protection structures	Provision of irrigation or potable water and hydroelectric power	Generate flooding elsewhere
Retrofitting schools	Continuation of schooling and the protection of school books	Use of schools as shelters may diminish their appropriateness as a suitable place for education
Improvements in civil society networks and linkages (to strengthen response capacity)	Improved governance and more organised social structures	Reduction in state responsibility for emergency response
Proper land-use and urban planning processes (to manage flood risk)	Delivery of basic necessities (e.g. potable water, drainage, sewerage, power and community facilities)	Reduction in tax revenues from businesses that would have located in high-risk places (e.g. coastal areas)
Emergency shelters	Community facilities (e.g. clinics or schools) in non-disaster periods	Investments in purpose-built schools and hospitals foregone
Improvements in water supply systems in rural areas (particularly those prone to drought and floods)	Water supply systems improved regardless of a disaster occurring	Water taken away from other places—increasing drought risk elsewhere
Construction and use of drainage pipes to reduce flood risk	Improved irrigation practices, possibly improved agricultural practices	
Community-based disaster preparedness	Improved women's involvement in community activities; enabling village men to continue to work their own land and be with their families	Reduction in state responsibility for coordinating preparedness
Installing more resilient wireless communications (particularly in hurricane-prone places)	Enhancing access to telephony and electronic data services	
Training farmers to diversify the use of crops and build resilience to drought	Reducing vulnerability to poverty	
Better monitoring of food supplies for drought early warning systems	Improving the food supply chain, possibly making it more cost-effective	

Source Adapted from DFID (2005) and White and Rorick (2010)

3.3 Challenges of Identifying Environmental and Socioeconomic Co-benefits

A key challenge in making DRM investment decisions is incorporating environmental and socioeconomic co-benefits into traditional economic assessments. As the previous section outlined, this is because of the shortcomings of using CBA approaches to assess these co-benefits, but it is also because of the idiosyncrasies of public decision-making with respect to public goods, such as disaster risk reduction. We describe these challenges in more detail below.

3.3.1 Shortcomings of Using Standard Cost–Benefit Analysis

Standard CBAs that aggregate costs and benefits face at least four major challenges that make it difficult to fully account for environmental and socioeconomic co-benefits associated with DRM.

First, environmental and social co-benefits most often cannot easily be monetised. How does one monetise the value of enhanced livelihoods? How does one value strengthened community empowerment, integrated environmental and social assessments with inclusive stakeholder consultation, improved environmental and social management systems and protected ecosystems and wildlife conservation? Because most such co-benefits are not easily monetised, they are not likely to be included in a traditional CBA. This will result in the aggregate benefits of a DRM investment—including direct benefits and co-benefits—being relatively undervalued compared with aggregate costs.

Second, many co-benefits, such as social cohesion and sustainable landscape management, are delivered beyond the life of a project. Thus, even if these benefits could be monetised, those occurring in the more distant future will be discounted significantly by any rate of interest chosen for an aggregate economic assessment. In present value terms, these co-benefits will then be relatively undervalued compared with costs that are more likely to occur in the near future. Once again, aggregate benefits will be undervalued compared with costs.

Third, the likelihood of a disaster occurring is, of course, uncertain, as is the timing of its occurrence. With both the likelihood and the timing of disasters being uncertain, decision-makers are unable to determine when a DRM investment should be taken. This means that any potential co-benefits arising from such an investment—irrespective of the occurrence of a disaster—are subject to deliberations over when and if such an investment decision should be made. Thus, potentially significant and more immediate co-benefits may be missed because of delays in investment decision-making.

Fourth, and perhaps most importantly, a traditional CBA does not easily account for the distributional aspects of DRM investments. CBA provides an aggregate assessment—that is, if aggregate benefits exceed aggregate costs, then the investment should be undertaken. Co-benefits, however, may address specific environmental and socioeconomic problems, such as those outlined in previous sections, and they are likely to have different impacts on the various groups, such as workers, local communities, indigenous peoples, investors and consumers. Critically, important distributional considerations are not included in a traditional aggregate economic assessment.

When decision-makers try to be more "business-like" and "efficient" in their investments in DRM, there is a risk that they might actually ignore important co-benefits that are critically important for their decision-making. In particular, understanding the distributional impacts of the investments with respect to specific social and environmental issues is crucial.

3.3.2 *Idiosyncrasies of Public Sector Decision-Making*

In addition to the shortcomings of CBAs, there are further idiosyncrasies of decision-making, especially with respect to investing in public goods such as DRM programmes, which hamper the mitigation of disaster risks and the generation of environmental and socioeconomic co-benefits.

First, political time horizons tend to be short. Leaders and ruling parties generally focus on more immediate issues. Potential co-benefits of DRM, on the other hand, may be longer in term and less visible to policy-makers. Thus they are likely to be of less interest to those currently in power and wanting to stay in power.

Second, there may be political risks in undertaking DRM projects, especially if critical constituencies perceive that the impacts may be potentially negative. For example, risk mitigation regarding environmental sensitivities within the Arctic Circle in Alaska will be off the agenda in the US when a highly contested mid-term election is approaching. Highlighting potential environmental and social co-benefits may not always be politically popular.

Third, the direct benefits of DRM are generally seen as a public good, or at least a good for a sizeable portion of the public. This means citizens may not see these benefits as a personal benefit to them and thus may not support political decision-makers who invest in them. Because of the economics of self-interest, public sector decision-makers—like their counterparts in the private sector—tend to underinvest in public goods. Thus, as we see below, making the environmental and social co-benefits visible may be critical to actually securing an investment in a DRM project.

Fourth, and perhaps most worrisome, public decision-making may be more attracted by investments that generate quick financials return for the Treasury or quick spurts in economic growth than by longer-term and less immediate investments to manage the risks of disaster in an environmentally and socially responsible way.

This preference for short-term growth rather than long-term sustainability and social equity may even encourage investments that increase the risk of disasters. Such investments could include promoting urban or industrial growth in sensitive areas such as floodplains, estuaries and coastal areas.

3.4 Opportunities to Promote Environmental and Socioeconomic Co-benefits

Although it might appear business-like and efficient to include all environmental and social impacts of their projects in a CBA, most companies and financial institutions today do not attempt to monetise all costs and benefits. Nor should governments. For some issues, such as worker safety, it is actually inappropriate to put an economic value on a company's efforts. For other issues, such as social cohesion or ecosystem integrity, an economic valuation makes little or no sense. Rather, the social and environmental impacts of projects—whether private or public—should be identified, assessed and managed. This is why expert credit agencies, commercial investment banks and others are using best practice environmental and social standards to address the potential impacts—both negative and positive—of public and private investment decisions. For DRM investments in particular, all three types of potential impacts highlighted in the Triple Dividend of Resilience framework need to be considered: the avoided losses, the economic potential unlocked by reducing background risk and the social, economic and environmental co-benefits.

3.4.1 Towards a DRM Co-benefits Framework

In the case of private sector investments, as can be seen in the framing of the International Finance Corporation (IFC) Performance Standards,[4] the emphasis is on reducing the environmental and social risks of a commercial project. In the case of public sector investments, however, such as DRM investments, the emphasis needs to be on ensuring net positive environmental and social co-benefits—both by mitigating any negative impacts and by enhancing any positive impacts.

In this respect, political decision-makers may want to refer to internationally accepted sustainability and safeguard frameworks from the IFC, the World Bank and others to establish a *co-benefits framework for DRM investments*. For example, the approach set out in IFC Performance Standard 1 on the assessment and management of environmental and social risks and impacts could be adopted for a DRM Co-Benefits Framework, as follows:

---

[4]http://www.ifc.org/performancestandards.

- Identify and evaluate DRM environmental and social impacts
- Mitigate potential co-risks—that is, avoid, minimise and compensate/offset
- Enhance potential co-benefits—that is, require, maximise and subsidise/integrate
- Manage DRM investments effectively to ensure net positive co-benefits
- Communicate, listen and respond to key stakeholders and
- Engage, disclose and disseminate to key stakeholders.

As discussed further below, further research on the co-benefits of existing DRM investments will also contribute to developing such a framework. This framework should also include a typology of co-benefits and co-costs that identifies key elements such as:

- Direct benefits versus co-benefits
- Direct costs versus co-costs
- Intended versus unintended co-benefits and co-costs
- Immediate versus long-term co-benefits and co-costs
- DRM-specific versus general public investment co-benefits and co-costs.

3.4.2 Qualitative Cost–Benefit Analysis

A practical step towards developing a robust environmental and socioeconomic framework for DRM investments could be to simply list direct benefits, co-benefits and costs. The list could include items such as the following:

Direct benefits of DRM

- Avoided direct disaster costs
- Avoided indirect disaster costs
- Avoided non-economic disaster impacts.

Co-benefits of DRM

- New, additional sources of income
- Improved labour and working conditions
- Efficient resource use and reduced pollution
- Enhanced community health, safety and security
- Responsible land acquisition and resettlement
- Conservation and sustainable use of biodiversity
- Respect for indigenous peoples
- Protection of cultural heritage.

Costs of DRM

- Planning
- Construction
- Labour
- Materials
- Opportunity costs of the allocation of resources.

Co-costs of DRM

- Increasing risk in non-target communities
- Increasing scarcities in other public goods and services
- Unintended changes in migration patterns
- Negative environmental and socioeconomic externalities.

In this respect, guidance from the US Federal Emergency Management Agency (FEMA) on a *qualitative* CBA may be useful (FEMA 2007). FEMA proposes a straightforward method called "simple listing", which identifies all cost and benefits of DRM actions and assigns priorities.

This kind of exercise can help decision-makers better understand the range of benefits—both direct benefits and co-benefits—of a potential investment decision. In so doing, the case for investing in DRM can be better communicated to stakeholders who are likely to be interested in the more immediate social and environmental co-benefits of a DRM programme or project than in the more distant and less transparent direct benefits of the project.

By showing that the investment has not only direct benefits in terms of reducing disaster losses, but also significant, visible and accountable co-benefits for society and for the environment, it will be easier to demonstrate that it provides more immediate returns in support of sustainable development. In short, an articulation of the co-benefits can make the case for investing in a DRM project.

3.4.3 Integrating DRM and Environmental Management

Different agencies have proposed a number of approaches and tools to link DRM and environment management planning processes. The Global Development Research Center (GDRC) provides an overview of tools for linking DRM to environmental management in an urban context, demonstrating how DRM investments can generate environmental co-benefits through associated process of environmental management.[5] It presents six tools for doing so, divided into three parts: planning options, assessment of options and implementing actions (see Table 3.2).

[5]http://www.gdrc.org/uem/disasters/disenvi/tools/.

Table 3.2 GDRC tools for integrating environment management and DRM

Parts	Tools
Planning options	Environmental Profiling (EP) Eco and Hazard Mapping
Assessment of options	Environmental Risk Assessment (ERA) Environmental Impact Assessment (EIA) Strategic Environmental Assessment
Implementing actions	Environmental Management System (EMS)

The overview of tools proposes creating "an environmental profile (EP) of an area or community" as a first step in understanding the interactions between the local environment and disaster vulnerabilities. An EP for DRM purposes provides valuable information for other environmental management issues and thus highlights environmental co-benefits. The overview explains:

> An "environmental profile" (EP) provides a systematic overview of the development, environment and disaster setting, and institutional arrangements of an urban area, which is designed to highlight the environment-disaster interactions, the critical issues, and the sectors and stakeholders directly concerned with them. In this tool, a brief introduction to the social, economic and physical features of the city is made.

Regarding EIAs, the overview proposes using these for DRM purposes to identify and generate important co-benefits:

> The EIA tool, well known for the assessment of development projects such as a dam or a complex of factory buildings, can also be used for disaster planning and management. Like the ERA tool, it enables informed decision making on the development of a city, but also in preparing for disasters (whether natural or man-made), and in monitoring and evaluation of the action taken.

Beyond planning, and in order to ensure continued implementation of measures to manage disaster risk and the environment, the overview suggests setting up an EMS as "a systematic way to ensure environmental issues are managed consistently and systematically". An EMS can ensure the integrated delivery of DRM and environmental and social co-benefits:

> Effectively applied, an EMS can help integrate environmental considerations within a larger disaster management plan.

The suggestions provided above demonstrate some ways in which a DRM Co-Benefits Framework could be set up using existing tools and processes, to integrate environmental and social management practices based on international and national best practice. This will help ensure DRM co-benefits are identified, assessed and enhanced in DRM investment processes. In so doing, the likelihood of ministries of finance approving DRM programmes and projects will also be strengthened.

3.5 The Need for DRM Co-benefits Case Studies

DRM case studies that highlight the benefits of investment in managing risk generally focus on the direct benefits. In some cases, they also focus on associated environmental and socioeconomic benefits that may arise in disaster response measures. Regarding co-benefits of DRM investments irrespective of whether there is a disaster, there is a need for more case studies. One opportunity is to revisit existing DRM cases and more thoroughly explore the co-benefits aspects of these cases. Another option is to research new cases with a specific focus on co-benefits.

3.5.1 Two New Case Studies

Here, we identify two new case studies, each illustrating some of the environmental and socioeconomic co-benefits described in previous sections. These co-benefits were not fully recognised until after the DRM investments were made so did not incentivise action. However, knowing that they exist could stimulate further action to reduce disaster risk. The cases need to be developed further, particularly if some of these elements are to be quantified and the size of the co-benefits is to be established for advocacy purposes.

Case study 1: Jamaica—the co-benefits of investing to reduce drought risk in agriculture

In Jamaica, the agriculture sector contributes about 6 % of gross domestic product (GDP) and employs 17–18 % of the labour force. Domestic agriculture is largely located on hillside plots with an average size of 1 acre with slopes above 15°; export agriculture (including coffee, banana, cacao and coconut) contributes to 22 % of total exports, raising $274 million in foreign exchange each year. A number of DRM programmes have focused on this sector, including the Jamaica Rural Economy and Ecosystems Adapted for Climate Change programme, which aims to strengthen local and national institutional capacity to support climate change adaptation and DRM within agriculture. The Planning Institute of Jamaica has also committed $9.9 million to the development and implementation of adaptation measures to strengthen agricultural productivity and coastal protection and build local capacity for natural resource management.

Of these investments, those that have focused on reducing drought risk in farming seem to offer particularly high potential for co-benefits. The installation of dedicated irrigation systems to overcome the impact of drought has helped farmers increase their productivity and output, as well as reducing soil erosion and deforestation by optimising previously inefficient farming practices (see Box 3.1).

Box 3.1: Resilience dividends of risk management in Jamaican agriculture

Key co-benefits of integrated DRM investments include the following:

1. Economic co-benefits: DRM irrigation projects have reduced the impact of droughts, particularly in Southern Clarendon and St. Elizabeth. These farming communities have also benefited from increased productivity and output relative to other areas, even in the face of drought over the April-June quarter in 2014 (Kellier 2014).
2. Social co-benefits: Training and shared learning on drip irrigation has strengthened social capital and built comradeship within the communities, especially among the farmers in the field.
3. Environmental co-benefits: A rainwater catchment tank and drip irrigation system in Lititz, St Elizabeth, has improved small-scale irrigation, resulting in higher yields, less soil erosion and deforestation and an increase in socioeconomic status.

These co-benefits have not yet been quantified—we do not know by how much productively has increased—but identifying them provides a useful first step in building the case for scaling up investment in drought risk management in Jamaica.

Case study 2: Mexico—the co-benefits of flood protection in urban areas

The World Bank and Mexico's Ministry of Finance elaborated a joint study to determine the impact of investment in flood defence in terms of reducing flood damage in the state of Tabasco between 2007 and 2010 (World Bank 2014). This found that the cost–benefit ratio of these investments was 4:1, contributing to avoided damages and losses when floods occurred in 2010 equivalent to $3 billion, or 7 % of the GDP of Tabasco. This figure does not capture the full range of co-benefits associated with these investments, however.

The capital of Tabasco, Villahermosa, has also seen improvements in the urban environment as a result of federal government investment in flood defence. Major DRM investments have stimulated local actors to take greater care of the environment. Small-scale projects with environmental benefits have been initiated, including tree planting on riverbanks to prevent landslides. People are beginning to dispose of litter more responsibly, throwing less in the streets or into drains to avoid these becoming blocked during rainy season.

A report by the Colegio de la Frontera Sur (Díaz-Perera 2013) suggests, however, that there have been unintended costs or negative externalities associated with the flood defence project in Tabasco. Channelling water away from the capital Villahermosa has led to increased flooding elsewhere in the state of Tabasco—mainly in rural areas. There have also been negative environmental impacts as a result of these large construction projects.

It seems floods are caused not only by heavy rainfall but also by the way the dams operate. A Universidad Nacional Autónoma de México study of the floods in 2007 stated, "These results suggest that the suitable operation of the dams, based on better forecasts, would have reduced considerably the damages caused by the event" (López-Méndez et al. 2008: 3). Overall, the Tabasco flood protection case study suggests better methodologies are needed to measure the full range of costs and benefits of DRM investments, including unintentional ones.

Together, these two case studies demonstrate a number of unexpected co-benefits associated with investment in DRM, including direct income-generating opportunities, environmental service co-benefits, increased social capital and behavioural changes.

3.5.2 Building on Existing Case Studies

Regarding revisiting existing cases, this section introduces World Bank and UN Development Programme (UNDP) cases that may be of interest.

Case study 3: Environmental and economic co-benefits from mangrove planting in Vietnam

A Vietnam Red Cross Mangrove Plantation and Disaster Risk Reduction project in the typhoon- and flood-prone coastal provinces of northern Vietnam has proven to have significant environmental co-benefits, including those related to carbon sequestration, nutrient retention, sediment retention, biodiversity habitats, flood attenuation, wastewater treatment and water supply and recharge. The 17-year project cost $8.88 million to set up and has involved the creation of 9,462 ha of forest (8,961 ha of mangroves) in 166 communes and the "protection of approximately 100 km of dyke lines" (IFRC 2011: 3). Table 3.3 demonstrates the estimated benefits and costs of this World Bank case study in selected communes from 1994 to 2025.

In terms of total benefits, including co-benefits, it is estimated that (IFRC 2011: 7):

- Approximately 350,000 beneficiaries have been reached directly and 2 million indirectly.
- There has been an "increase in per hectare yield of aqua culture products such as shells and oyster by 209–789 %".
- Economic benefits from aqua product collection, honeybee farming, etc., are found to be between $344,000 and $6.7 million in the selected communes.
- Environmental benefits include $218 million in terms of estimated minimum CO_2 emissions absorbed by the planted mangroves (assuming a price of \$20/tCO_2e).

Table 3.3 Estimated benefits and costs in selected communes of Vietnam, 1994–2025

Commune	Dai Hop	Thai Do	Nam Thinh	Giao An	Dien Bich
District	Kien Thuy	Thai Thuy	Tien Hai	Giao Thuy	Dien Chau
Province	Hai Phong	Thai Binh	Thai Binh	Nam Dinh	Nghe An
Population	10,955	6,087	7,240	10,496	10,521
Sea coastline (km)	3.9	5.5	5.9	3.2	3.5
Dyke line (km)	4.0	7.5	5.9	3.2	3.5
Timeframe of planting	1998–2005	1994–2005	1997–2005	1997–2005	1998–2005
Planting input (ha)	835	1,010	1,287	2,403	145
Planting output (ha)	450	900	380	678	100
Planting-related costs ($)	425,866	858,373	362,424	646,641	95,374
Protective benefits ($)	676,868[c]	15,330,243	N/A[d]	37,818,545	N/A[e]
Direct economic benefits ($)	628,094	672,436	4,799,476	6,748,533	344,931
Ecological benefits ($)	10,989,000	32,730,828	12,307,055	23,308,814	3,437,879
Total identified benefits ($)	12,293,962	48,733,507	17,106,531	68.375,892	3,782,810
Benefit/cost ratio 1[a]	3.06	18.64	13.24	68.92	3.61
Benefit/cost ratio 2[b]	28.86	56.77	47.20	104.96	39.66

Notes [a]Excludes ecological benefits. [b]Includes ecological benefits. [c]Protective benefit concerns only reduced damages to the sea dyke. [d]Protective benefits were identified but could not be attributed to the project. [e]Protective benefits were identified but could not be quantified
Source IFRC (2011)

Case study 4: Economic and livelihood co-benefits in Lao People's Democratic Republic

The World Bank Mekong Integrated Water Resources Management Project has produced a number of economic co-benefits. Estimates of these are based on the assumption that full benefits would be reached in three years and that the "economic life of the investment is 15 years" (World Bank 2012: 117). The project was designed with water resource management and floodplain management at its core and resulted in the rehabilitation of 10 floodgates in the Xebangfai River and about 40 village irrigation schemes in the Xebangfai and Xebanghieng Rivers of Lao PDR. The floodgate rehabilitation increased flood protection, but it also resulted in increased agricultural production (especially rice), increased fish catch in

Table 3.4 Summary of floodgate rehabilitation activities

	Required works	Estimated financial cost ($)	Estimated economic cost ($)	Benefit area (ha)	Flood protection benefit ($)	Estimated fish benefit ($)	Total benefit ($)
Huay Pin	Rehabilitation of mechanical works (gates), minor structural repairs to headworks	72,000	68,400	120	12,375	3600	15,975
Huay Kae	Rehabilitation of mechanical works (gates), minor structural repairs to headworks	52,500	49,875	100	10,313	3600	13,913
Huay Pa Pak	Rehabilitation of mechanical works (gates), minor structural repairs to headworks	35,000	33,250	100	10,313	3600	13,913
Huay Bung Or	Rehabilitation of mechanical works (gates), resectioning of canal (2.5 km)	31,875	30,281	150	15,469	3,600	19,069
Huay Daeng	Rehabilitation of mechanical works (gates), resectioning of canal (3 km)	38,250	36,338	170	17,531	3,600	21,131
Total			218,144	640	66,000	18,000	84,000

Source World Bank (2012)

the floodplain and decreased electricity costs through increases in water use efficiency (World Bank 2012). Table 3.4 presents a summary of the floodgate rehabilitation activities and estimated co-benefits.

Another component of the project included fisheries management, which aimed to support the rural infrastructure and rehabilitation/construction of hatcheries. Although no prior economic analysis was possible, "experience from similar projects in the region suggests that the rural infrastructure identified by the communities would have significant economic returns … reflect[ing] the priorities of the communities" (World Bank 2012: 116), and demonstrating co-benefits in terms of livelihood opportunities and returns.

Case study 5: Social co-benefits of women's secure housing programme in Philippines
Quantifying social co-benefits such as community empowerment and social inclusion is difficult. Engaging grassroots organisations in the design and implementation of DRM initiatives, for instance, has proven extremely beneficial but difficult to quantify. In the Philippines, the Damayan ng Maralitang Pilipinong Api (DAMPA) women's group has been working on a project to help secure housing and livelihoods for the urban poor in Metro Manila, an area prone to both earthquakes and floods. In addition to helping 3500 poor families secure land and housing tenure, partnerships have been built with the government, civil society and private agencies that improve service delivery to poor communities, thereby making the initiatives more sustainable.

This UNDP project has provided DAMPA with improved access to "knowledge, information, financial resources and ability to advocate for government programmes that are responsive and accountable to the urban poor" (Fordham and Gupta 2011). In addition, volunteers are now sitting on "various local development and planning bodies organized at the local government level, where they actively participate in local governance and development planning activities in their localities" (ibid.), meaning that can continue to advocate for the needs and voices of the poor.

Overall, despite the examples and suggested co-benefits highlighted in this chapter, a comprehensive co-benefits framework is needed that includes and categorises all potential environmental and socioeconomic impacts associated with DRM. Co-benefits research focused on revisiting existing case studies or developing new case studies could play an important role in this regard.

3.6 Conclusions and Next Steps

This chapter demonstrates that a business-as-usual approach to DRM is unlikely to focus sufficiently on the potential environmental and socioeconomic co-benefits of DRM. This is because these benefits either are going unnoticed or are not seen as important to standard decision-making processes for DRM—in part because they are not included in traditional CBAs. Nevertheless, co-benefits are already being generated by existing DRM projects and further case study research could help us obtain a better understanding of these hidden benefits.

The development and testing of a DRM Co-Benefits Framework could help decision-makers focus on the potential co-benefits of programmes and projects aimed at building disaster resilience. Highlighting these would not only benefit the natural environment and affected communities but also enhance the "business case" for DRM and thus facilitate effective decision-making. In this respect, currently accepted environmental and social standards for investment and qualitative approaches to CBA may prove useful starting points. Within such a framework, as appropriate, the use of reliable, quantitative CBA information will also need to be included.

A good place to begin might be further research on current internationally funded DRM projects that have integrated environmental and social safeguards to avoid negative impacts. As these projects are explicitly addressing environmental and social impacts, they could offer substantive insights on the potential for the DRM investments to also generate positive environmental and social co-benefits—irrespective of whether a disaster occurs.

References

De Villiers B (1999). Disasters: the Social impact. Paper presented at the DMISA conference, Sea Point, Cape Town, 2–3 Sept 1999

DFID (Department for International Development) (2005) Natural disaster and disaster risk reduction measures: a desk review of costs and benefits. Draft Report. DFID, London

Díaz-Perera MÁ (2013) La construcción de las condiciones históricas de posibilidad de un desastre: el caso de dos colonias de Villahermosa, Tabasco. Final report of FORDECYT Cuenca Grijalva Project. El Colegio de la Frontera Sur, Villahermosa, Mexico

FEMA (Federal Emergency Management Agency) (2007) *Using* benefit-cost review in mitigation planning state and local mitigation planning. How-To Guide 5. FEMA, Washington, DC

Fordham M, Gupta S (2011) Leading resilient development. Grassroots women's priorities, practices and innovations. Groots International, Northumbria University School of the Built and Natural Environment and UNDP, New York

IFRC (International Federation of Red Cross and Red Crescent Societies) (2011) Mangrove Plantation in Viet Nam: measuring impact and cost benefit. IFRC, Geneva

Kellier D (2014) Statement on drought conditions and impact on the agricultural sector. Address, 29 July

López-Méndez V, Zavala-Hidalgo J, Romero-Centeno R, Fernández-Eguiarte A (2008) Analysis of the extreme flooding during October 2007 in Tabasco, Mexico using the WRF Model. Paper presented at the 9th WRF users' workshop, National Center for Atmospheric Research, June 23–27, 2008

Tanner T, Surminski S, Wilkinson E, Reid R, Rentschler J, Rajput S (2015) The Triple Dividend of Resilience: realising development goals through the multiple benefits of disaster risk management. ODI and World Bank, London

White Bo, Rorick Mary (2010) Cost-benefit analysis for community-based disaster risk reduction in Kailali. Mercy Corps Nepal, Lalitpur

World Bank (2012) Proposed Adaptable program lending (grants) for a mekong integrated water resources management project. Project Appraisal Document, 8 February. World Bank, Washington, DC

World Bank (2014) Análisis de los impactos de las inversiones en prevención y reducción de riesgos: estudio de caso de Tabasco entre 2007 y 2010". Not published

Chapter 4
Disaster Risk Management and Fiscal Policy: Entry Points for Finance Ministries

Reinhard Mechler, Junko Mochizuki and Stefan Hochrainer-Stigler

Abstract This chapter reflects on the benefits of disaster risk management (DRM) in the context of fiscal policy and public investment. Of particular interest is the question of how those in charge of fiscal policy decisions can recognise and realise the economic and broader benefits of DRM. We consider the interplay between public DRM investment and fiscal policy and provide an overview of current debate as well as assessment methods, tools and policy options. Standard practice has been to focus on direct liabilities and recurrent spending, dealing the costs of disasters often only after the fact. Their full costs have thus often not been budgeted for; with a price signal missing, there is lack of clear incentives for investing in DRM.The discussion traces progress by focusing strongly on analytics and current practice. Overall, we identify four steps, being pursued deliberately: (1) assessing the relevance of disaster risk for public finance; (2) protecting public finance through risk-financing—examining insurance-related instruments that support protection of the fiscal position (first dividend of resilience); (3) comprehensively managing disaster risk, including reduction and preparedness as they affect development (second dividend of resilience); and (4) pursuing a synergistic co-benefits strategy of concurrently managing disaster risks and promoting development (third dividend of resilience).

Keywords Sovereign risk · Contingent liabilities · Fiscal policy · Risk financing · Insurance

---

The authors are very grateful for substantial guidance and very helpful comments received from Swenja Surminski, Thomas Tanner, Stephane Hallegatte and Mook Bangalore.

R. Mechler (✉) · J. Mochizuki · S. Hochrainer-Stigler
International Institute for Applied Systems Analysis, Laxenburg, Austria
e-mail: mechler@iiasa.ac.at

J. Mochizuki
e-mail: mochizuk@iiasa.ac.at

S. Hochrainer-Stigler
e-mail: hochrain@iiasa.ac.at

S. Surminski and T. Tanner (eds.), *Realising the 'Triple Dividend of Resilience'*, Climate Risk Management, Policy and Governance,
DOI 10.1007/978-3-319-40694-7_4

4.1 Introduction: From Understanding Risk to Building Fiscal Resilience

4.1.1 Overview

Disaster risk has seen strongly increased recognition in research, policy and implementation over the past few years. Substantial investments in disaster risk management (DRM) have been made and, according to some accounts, the balance between wait- and see (*ex-post* relief and reconstruction funding) and pro-action (*ex-ante* investments in DRM) has shifted from 95 versus 5 % a decade back to about 87 versus 13 % (Kellett and Caravani 2013). Economic decision support tools have helped us understand the benefits of DRM and shown substantial dividends (see UK GSO 2012).

Yet more effort is needed to further shift this balance. Recent UN Office for Disaster Risk Reduction (UNISDR) Global Assessment reports (UNISDR 2013, 2015a) issue a stark warning that economic losses linked to disasters are "out of control" and will continue to escalate unless disaster risk management becomes a core part of business investment strategies. The World Bank's 2014 World Development Report (World Bank 2013) emphasises the need to further switch from unplanned and ad hoc responses to proactive and systematic risk management. In addition, recent Intergovernmental Panel on Climate Change (IPCC) assessment reports (Field et al. 2012, 2014) emphasise the need for risk-based assessment and careful management planning before disasters strike. Finally, the last Global Risk Report published by the World Economic Forum (WEF 2015) concludes that stronger efforts are needed to understand, measure and foresee the evolution of interdependencies of risk.

Governments at different scales are important actors in DRM. In addition to providing DRM, regulating private sector activity and acting as promoters and coordinators of collective action on DRM (Wilkinson 2012), they are risk-takers, as a large part of disaster risk ends up with the fiscal position (Mechler 2004). Over the past few years there has been increasing recognition and understanding of the need to deliberately consider this in public and fiscal risk planning for disasters and implement DRM to the extent possible.

4.1.2 Approach

This chapter reflects on the benefits of DRM in the context of fiscal policy and public investment, addressing the question of whether and how co-benefits through disaster resilience-building can be further promoted. In line with the literature, we define co-benefits as positive externalities that arise deliberately as a result of a joint strategy that pursues several objective synergistically at the same time, such as

DRM and development goals or DRM and climate change adaptation (see Hourcade and Shukla 2001).[1]

Of particular interest for the following debate is the question of how those in charge of fiscal policy decisions can recognise and realise the economic and broader benefits of DRM. The discussion considers the interplay between public DRM investment and fiscal policy and provides an overview of current debate as well as assessment methods, tools and policy options. Currently, in fiscal budgeting practice, it is mostly standard to focus on direct liabilities and recurrent spending, such as foreign and domestic sovereign borrowing, expenditures by budget law, future recurrent costs of public investment projects and pension and health care expenditure. Costs of disasters are often dealt with after the fact only, rather than being considered contingent liabilities. As a consequence, the full costs of disasters are not budgeted for and, with a price signal missing, there is lack of clear incentives for investing in DRM.

4.1.3 Charting Out Progress

The following discussion traces progress in the debate on fiscal disaster risk management by focussing strongly on analytics and current practice (Fig. 4.1). We identify four steps, which are being pursued deliberately, as well as three dividends, which are being harnessed (Tanner et al. 2015).

1. Understanding fiscal risk—identifying and assessing the relevance of disaster risk for public finance
2. Protecting public finance through risk-financing instruments—identifying and examining insurance-related instruments that support protection of the fiscal position (first dividend)
3. Comprehensively managing disaster risk, including risk reduction and risk preparedness as they affect development (second dividend)
4. Pursuing a synergistic co-benefits strategy of concurrently managing disaster risks and promoting development (third dividend).

Specifically, this chapter aims at providing an analytical assessment with a "user focus" based on the following broad guiding question: How can the findings support government's DRM investment decisions as a public good? We provide empirical evidence, seek to identify good/bad practices in fiscal policy design and contextualise the discussion with relevant country-level and regional examples, such as from Mexico, the Caribbean states and Organisation for Economic Co-operation and Development (OECD) countries. Overall, we seek to distil entry

[1]In contrast, ancillary benefits are additional benefits that arise without deliberate planning. Similarly, there may also be co-costs from projects and policies. This is not the topic of this chapter, but will need attention further on.

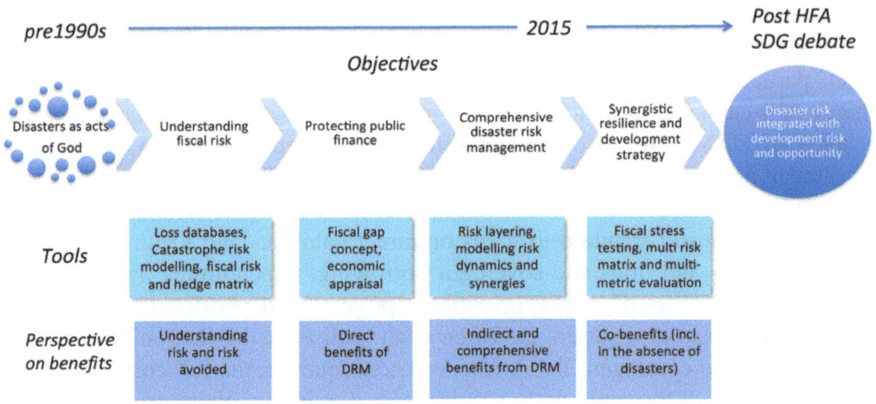

Fig. 4.1 Tracing progress in debate and practice—from acts of god to DRM as part of managing development risk. *Source* The authors

points for more strongly recognising and realising the economic and broader benefits of DRM for those in charge of fiscal policy. Specifically, we identify current guiding principles and aims for fiscal policy, then discuss whether and how these can be amended to support DRM.

The ensuing discussion is organised according to the four steps and dividends and then leads into a short conclusions section (Sect. 4.6), which provides a final commentary regarding the on-going transition, which increasingly positions disaster risk as part and parcel of resilience strategies to harness co-benefits from managing disaster risk and stimulating development.

4.2 Understanding Fiscal Risk

A first logical, and commonly pursued, step in managing (fiscal) risk is to properly understand and put such risk in the context of fiscal operations. Considerable effort has been expended on this over the past few years.

Natural disasters lead to loss of life and assets and have large impacts on people, businesses and governments. Governments at different scales are key to assessing, reducing and financing disaster risk. From an economic perspective, they are exposed to natural disaster risk and potential losses through three functions: (1) the allocation of goods and services (security, education, clean environment); (2) the provision of support to private households and business in the case of market failure; and (3) and the distribution of income (Mechler 2004; Musgrave 1959). From a budgeting perspective, sovereign disaster risk arises as a contingent public sector liability, which is associated with government's functions to provide relief, support recovery, undertake reconstruction and raise tax revenue. Once a disaster hits, these contingent liabilities can lead to large costs to governments in relation to providing relief, recovery and reconstruction assistance (see Box 4.1).

Box 4.1: Government operations and costs post-disaster

Relief operations include emergency assistance provided to the affected population to meet basic needs, such as shelter, food and medical attention.

Early recovery operations following the initial relief efforts are crucial to limit secondary losses and ensure reconstruction can start promptly. They include the emergency restoration of lifeline infrastructure (e.g. water, electricity and transportation lines), the removal of debris and the like.

Reconstruction operations generally centre on the rehabilitation or replacement of assets damaged by a disaster. These include public facilities and infrastructure, which are the direct responsibility of the state, but national or municipal authorities usually face obligations that go beyond their own assets. Governments are often called on to subsidise the reconstruction of private assets, in particular housing for low-income families that could not otherwise afford to rebuild their homes.

Loss of tax revenue arises as the economy is depressed and needs time to recover.

Source: World Bank (2010a).

4.2.1 Coping with Risk: Understanding Risk Tolerance and the Need to Plan

As disaster risk is a liability, a question arises as to whether governments should take it into account explicitly or whether it can afford a responsive mode of operations. A seminal paper by Arrow and Lind (1970) on the role of sovereign risk preference proposed that governments behave (disaster) risk neutrally, as they are considered the entity best suited to deal with risk via efficiently pooling and spreading potential losses. More precisely, the argument did not favour neglecting risk; rather, Arrow and Lind suggested a fiscal management approach based on expected values only: "The government should behave as an expected-value decision maker" (1970: 366). This means governments, as they can afford to refinance quickly, should plan and reserve only for average costs incurred over longer time horizons, and do not need to pay close attention to variability in costs, which arises because disasters are high-impact low-frequency events (and thus are defined by strong volatility around the mean).

Over the past few years, however, it has been recognised that variability matters and that countries exhibit differential coping capacities for dealing with risk (see Box 4.2).

Box 4.2: Understanding risk preference—the case of US and Haiti

Figure 4.2 contrasts recent large disasters in two very countries with extremely different coping capacities: the US and Haiti. In the US, Hurricane Katrina caused colossal losses of about $125 billion, but this amounted to less than 1 % of gross domestic product (GDP). While absolute losses for Haiti were smaller, in relative terms they were tremendous, at more than 160 % of GDP. Serious negative fiscal and macroeconomic effects have to be expected in the medium to longer term, although in practice these effects are often not monitored and are difficult to isolate from the background noise (see Noy 2009). However, comprehensively spreading the losses using tax revenue or savings seems impossible for Haiti. This owes partly to its smaller population and small total area as well as to its relatively low tax revenues in terms of GDP.

Practically speaking, the Arrow-Lind theorem has been challenged on theoretial grounds and the case for risk aversion has been understood (Anginer et al. 2013; Ghesquiere and Mahul 2007; Hochrainer 2006; Mechler 2004; Mechler and Hochrainer-Stigler 2014; Priest 2003). However, few of these analyses (Ghesquiere and Mahul 2007; Hochrainer 2006; Mechler 2004; Mechler and Hochrainer-Stigler 2014) explicitly study and criticise the details of the theorem for the disaster dimension. Broadly, the Arrow-Lind theorem does not apply to governments of countries that exhibit some of the following characteristics (Mechler 2004; Mechler and Hochrainer-Stigler 2014), and in these cases governments should justifiably act as risk-averse agents.

- High natural hazard exposure
- Economic activity clustered in a limited number of areas with key public infrastructure exposed to natural hazards and
- Constraints on resources to finance disaster losses and associated requirements. Such sources are determined by ability to reallocate the budget, domestic savings, access to financial markets and level of external indebtedness.

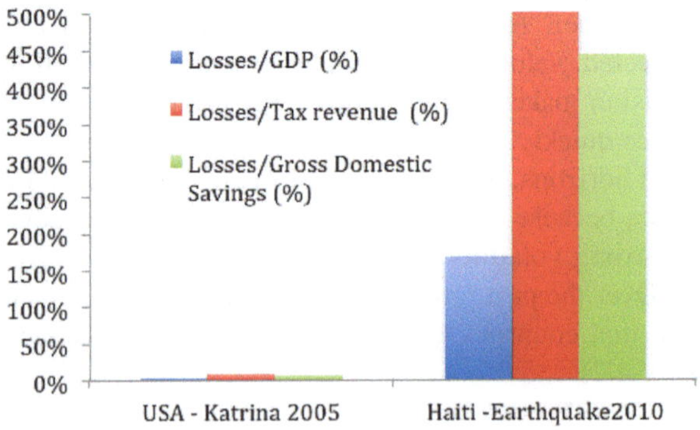

Fig. 4.2 Differential ability to spread risk for two large disasters, in the US and Haiti. *Source* Mechler and Hochrainer-Stigler (2014)

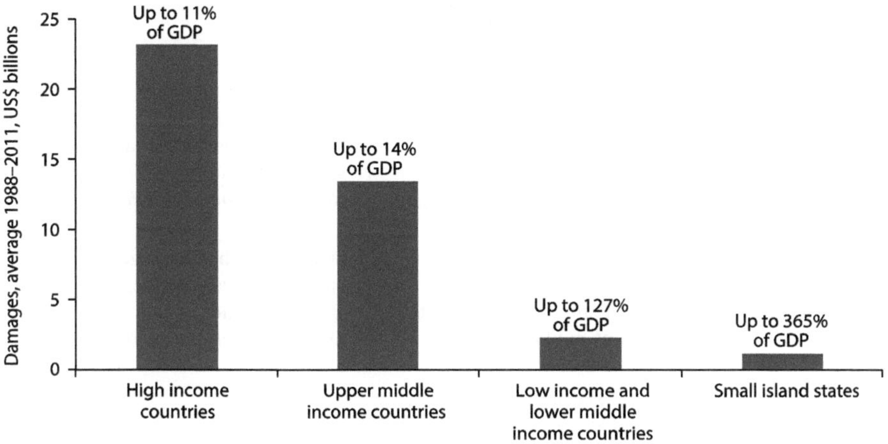

Fig. 4.3 Country income groups and disaster losses. *Source* World Bank (2013)

While income is not the sole defining variable for risk coping and risk preference, it is informative to compare income with losses in large events to understand where to look and where to prioritise action. As Fig. 4.3 suggests, while absolute damages (losses) hve been concentrated in higher-income countries, in lower-income countries and particularly in small island developing states (SIDS) the relative burdens have been found to be much larger [e.g. more than 300 % of GDP in SIDS (World Bank 2013)].

If risk aversion is identified as the proper risk preference, this implies that risk has to be taken into account explicitly (and beyond average values) in budget and fiscal planning. A government then should go beyond being an "expected value" decision-maker and consider variability and risk properly.

4.2.2 Tools and Concepts: The Fiscal Risk and Hedge Matrices

Budget and resource planning for disasters is not an easy proposition. Applying the so-called "fiscal risk matrix" is a step forward. Governments commonly plan and budget for *direct* liabilities—that is, liabilities that manifest themselves through certain and annually recurrent expenditure. Those liabilities are termed "explicit" (as recognised by law or contract) or "implicit" (moral obligations). In contrast, disaster risk enters the balance sheet as *contingent* liabilities (marked in italics in Table 4.1)—that is, obligations that arise randomly when a particular event occurs. Explicit, contingent liabilities deal with the reconstruction of infrastructure destroyed by events, whereas implicit obligations are associated with providing

Table 4.1 Government liabilities—the fiscal risk matrix

Liabilities	**Direct** Obligation in any event	**Contingent** Obligation if a particular event occurs
Explicit Government liability recognised by law or contract	• Foreign and domestic sovereign borrowing • Expenditures by budget law and budget expenditures	• State guarantees for non-sovereign borrowing and public and private sector entities • *Reconstruction of public assets*
Implicit "Moral" obligation of the government	• Pension and health care expenditure • Future recurrent costs of public investment projects	• Default of subnational government or public or private entities • Banking failure • *Disaster relief and recovery assistance*

Note DRM-relevant items marked in italics
Source Adapted from Polackova Brixi and Mody (2002)

relief—commonly considered a moral liability for governments (Polackova Brixi and Mody 2002).

Similarly to the fiscal risk matrix, a fiscal hedge matrix can be established, which would identify the sources governments have available to generate resources generally and in future (contingent) events (Table 4.2). Risk-financing would fall under the explicit contingent sources for coping with disaster losses.

Three key types of government risk-financing are worth noting in relation to disasters (and are discussed further below): reserve funds, contingent credit lines and sovereign insurance (traditional or alternative).

Table 4.2 Government sources—the fiscal hedge matrix

Liabilities	**Direct** Sources in any event	**Contingent** Sources if a particular event occurs
Explicit Direct control by government	• Tax revenues (less tax expenditures) • Government-owned assets for possible sale or lease	• Transfer income from central government • Recovery of loans made by government (on-lending) • Legal claims against the state • **Reserve funds** • **Contingent credit lines and financing commitments from official creditors** • **Sovereign insurance**
Implicit Not directly controlled by government	Existing funds that are under indirect government control (social security funds)	Future profits of state-owned enterprises and agencies

Note DRM-relevant items in bold
Source Adapted from Polackova Brixi and Mody (2002)

4.2.3 Evidence of Planning for Contingent Liabilities

Historically, countries have generally not planned for contingent liabilities, and *ex-post* sources such as budget reallocations, aid and emergency loans have financed disaster losses. Large developed countries have relied on national reserve funds, reallocation of the budget (existing tax revenue) or new tax revenue to fund the aftermath of disaster events, and have done less fiscal planning for disaster risk (see, e.g., UNESCAP 2013). OECD and larger countries can generally absorb the impact of adverse natural events since revenues from unaffected regions can subsidise the affected region.

The fiscal risk matrix has seen application with reference to disaster risk in Colombia, Indonesia, Mexico and Thailand, among others. Colombia has been one of the pioneers in this regard (see Box 4.3).

Box 4.3: Assessing the contingent liability of disasters using catastrophe risk models in Colombia
Colombia is a leader in assessing contingent liabilities. In 2010, the government for the first time undertook a comprehensive assessment of all such liabilities. Natural disaster risk was found to be the second most important liability (after legal claims on the state, which ranked top), with annual expected losses estimated at close to $0.5 billion or 0.7 % of the 2010 budget. While annual averages are informative, variability in terms of low-frequency, high impact events is key for this discussion: 100-, 250- and 500-year return period events were considered to potentially lead to losses amounting to about 4, 7 and 8 % of budget resources, respectively (Table 4.3).

Table 4.3 Contingent liabilities assessment for Colombia

Contingent liability	$ million	% of GDP	% of budget
Legal actions	18,642	7.5	27.7
Natural disasters	490	0.2	0.7
Public credit operations	56	0.01	0.1
Infrastructure projects	26	0.01	0.0
Probable maximum loss (PML) from natural disasters			
100-year PML	2976	1.2	4.4
250-year PML	4417	1.8	6.6
500-year PML	5655	2.3	8.4

Source Ministry of Finance and Public Credit, Colombia (2011), in GFDRR (2012b)

4.2.4 Fiscal Stress Testing

Improved understanding of risk has been the basis for fiscal stress testing, for which decision-supporting tools have been developed. As an indicator of financial vulnerability, Mechler (2004) suggests measuring sovereign financial vulnerability in terms of the "resource gap" concept. Accordingly, in terms of lack of sufficient funding for relief and reconstruction, governments would be fiscally risk-averse if they could not access sufficient funding after a disaster to cover their liabilities with regard to reconstructing public infrastructure and providing assistance to households and businesses. The repercussions of large resource gaps can be substantial. The inability of a government to repair infrastructure in a timely manner and provide adequate support to low-income households can result in adverse long-term socioeconomic impacts. As a case in point, despite substantial inflows of donor aid, but given limited domestic resources, Honduras received only about 50 % of the funds necessary for relief and reconstruction, and experienced extreme difficulties in repairing public infrastructure and assisting the recovery of the private sector following Hurricane Mitch in 1998. Five years after Mitch's devastation, the GDP of Honduras was 6 % below pre-disaster projections.

A report by the World Bank (Cummins and Mahul 2008) adds another dimension to this framing and assessment in terms of the timing of resource flows. While enough funding may be available over time, there may be a sporadic resource gap. This is because in the aftermath of a disaster event urgent expenditure needs are generally high, but the immediately available financial resources are often very limited. The timing of financial inflows for financing the losses is important and can differ for different *ex-ante* and *ex-post* instruments.

Although there has been a considerable amount of discussion, there is very little reported evidence on the scope and scale of liquidity gaps. The case of Grenada is a notable exception, highlighting as it does various repercussions of fiscal crisis (see Box 4.4).

Box 4.4: Grenada and the financing gap post-Hurricane Ivan

Hurricane Ivan struck Grenada on 7 September 2004 and left tremendous devastation in its wake, with damages estimated at over $800 million—or twice Grenada's GDP. Just as it required additional resources to finance relief, the clean-up and emergency rehabilitation, Grenada experienced a dramatic decline in revenues—an estimated 5 % of GDP between September and December 2004. The government, which had only limited reserves, faced serious problems financing the public service bill, including salaries and the continuation of key services. It also became evident that the country would not be able to meet its debt obligations as they fell due.

In an effort to secure the necessary resources to continuing functioning, the government sought donor assistance in the reconstruction of the island and in helping meet its expense liabilities (imports and civil servant salaries).

Despite over $150 million in pledges, only $12 million was available to address immediate liquidity needs. The remainder of the funds pledged was earmarked for reconstruction projects implemented over the following two years. The government also sought the cooperation of its creditors by developing a proposal to restructure over 85 % of its commercial debt. Its final effort to address its revenue shortfall was to pass revenue-enhancing measures yielding over 2 % of GDP in April 2005, about seven months after the event. These measures included (1) an increase of about 45 % in the retail price of fuel; (2) an increase in excise taxes on alcohol and tobacco; (3) a special levy on incomes over $375 per month for a five-year period; and (4) improved tax administration.

Despite all these efforts, Grenada's fiscal situation remained challenging and the country still faced a financing gap of 4.5 % of GDP for 2005, with total debt projected to increase to 150 % of GDP. Furthermore, instead of focusing on recovery and reconstruction, the government was distracted by the need to finance the emerging resource gap. This led to delays in recovery and reconstruction. To make matters worse, Hurricane Emily followed in 2005, causing about $50 million in additional economic losses. The Grenada experience and lessons learnt are considered to have been an important impetus for discussions regarding the creation of the Caribbean Catastrophe Reinsurance Facility (CCRIF) in 2007 (see World Bank 2010b).

However, the detailed information available on Grenada, including a reported instance of a liquidity gap, is the exception rather than the rule. At best the information available is often fragmentary.

4.2.5 Analytical Tools to Assess Fiscal Risk and Gaps

Given lack of robustness of empirical information, interested parties may want to resort to analytical tools to derive relevant information. Modelling and decision support based on work by the International Institute for Applied Systems Analysis (IIASA) (see Mechler 2004; Hochrainer 2006; Hochrainer-Stigler et al. 2014; IADB 2008) over the past few years is available regarding countries' financial vulnerability and questions relating to how much and what to insure. The Catastrophe Simulation (CATSIM) model, developed by IIASA, has addressed this question in some detail for many countries and regions. CATSIM is a risk-based economic framework for evaluating economic disaster impacts and the costs and benefits of measures for reducing those impacts. CATSIM uses stochastic simulation of disaster risks by randomly and repeatedly generating disaster events in a

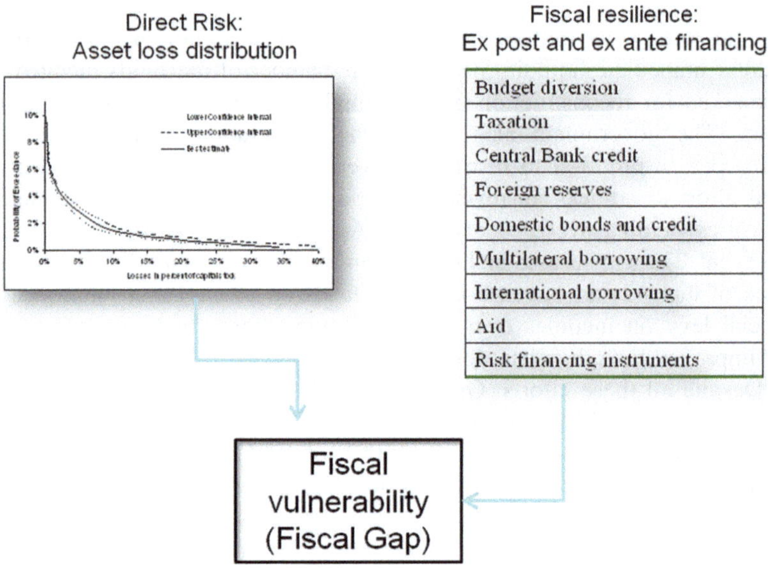

Fig. 4.4 Modelling fiscal vulnerability and resilience to natural hazards. *Source* Hochrainer-Stigler et al. (2014)

specified region and examines the ability of the government and the private sector to finance relief and recovery. The model compares asset loss distribution with fiscal resilience, defined as the total of *ex-post* and *ex-ante* risk-financing (see Fig. 4.4).

For the 2010 World Development Report and Hochrainer-Stigler et al. (2014), CATSIM was used to conduct global analysis on fiscal vulnerability and risk. This highlighted the following countries as particularly fiscally vulnerable: (1) various SIDS in the Caribbean and Pacific; (2) countries in Latin America (Bolivia, El Salvador, Honduras, Nicaragua), Africa (Madagascar, Mauritania, Mozambique, Nigeria, Sudan, Zimbabwe) and Asia (Cambodia, Indonesia, Laos, Nepal, Papua New Guinea, the Philippines). These countries are prime candidates for stepping up activities to plan, reduce and manage risks in order to reduce the serious human and financial loss burden for exposed populations, businesses and the wider macro-economy. Figure 4.5 shows a global map of fiscal gap return periods—that is, the estimated return period for which countries would incur a fiscal shortfall.

The Disaster Deficit Index (DDI) developed by the Inter-American Development Bank (IADB 2008) is based on the CATSIM methodology and can be derived by dividing the loss by the financing available. For example, in Fig. 4.6, the DDI of about 4.3 for a 50-year return period event in Honduras means losses would amount to more than four times the finance available to rebuild lost assets.

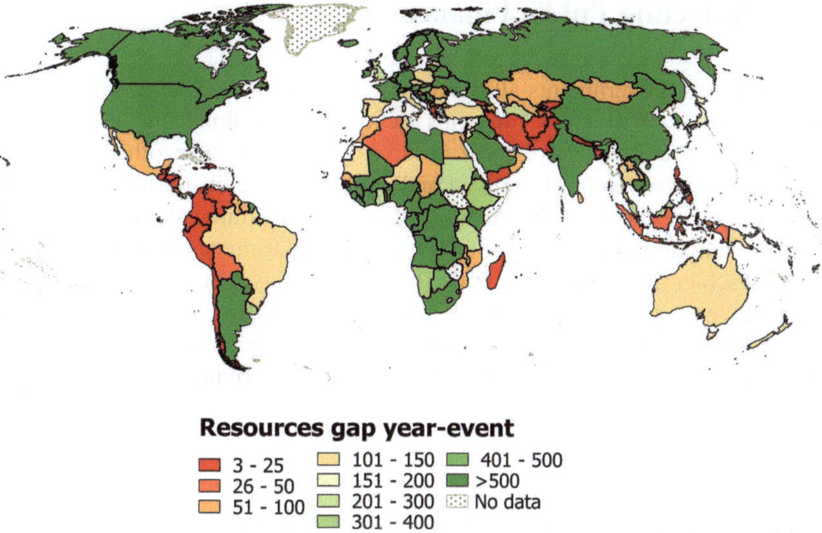

Fig. 4.5 Global map exhibiting calculations of the fiscal gap year. *Source* Based on Williges et al. (2015)

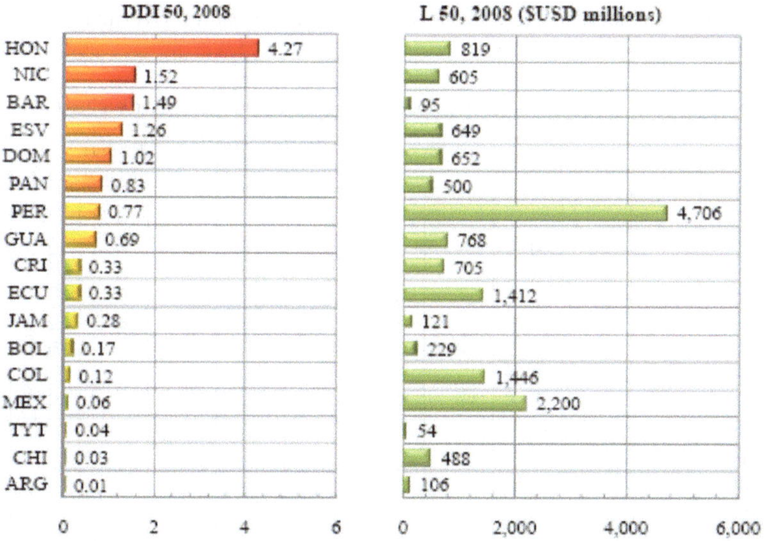

Fig. 4.6 Calculating the DDI for Latin American and the Caribbean—DDI and PML in 50 years. *Source* IADB (2008)

4.3 Protecting Public Finance

As mentioned, governments all over the world have relied on *ex-post* resources to fund the costs of disasters. In terms of *ex-ante* risk-financing instruments, government reserve funds have been used; only recently have other *ex-ante* instruments, such as sovereign insurance and contingent credit, started to be employed. Increasingly, because of the delays and uncertain timing of *ex-post* instruments, attention has turned to the use of such *ex-ante* instruments (Cummins and Mahul 2008; Mechler 2004; Gurenko 2004; Linnerooth-Bayer et al. 2005; OECD 2012). Over the years, given identified risk aversion, fiscal gaps, their size and timing, *ex-ante* risk-financing instruments, such as insurance, reserve funds and contingent credit, have been considered to complement the commonly employed *ex-post* instruments.

4.3.1 *Risk-Financing and Planning Practice*

A number of countries that are highly susceptible to disaster risk have begun to consider disaster and budget planning and to move more strongly from a reactive to a proactive perspective. These include Colombia, Mexico, Caribbean countries and Pacific SIDS. In Asia also there is momentum in terms of a movement from *ex-post*

Table 4.4 Summary of fiscal risk management arrangements in ASEAN member countries

	Ex-post						*Ex-ante*				
									Risk transfer		
	Annual budget allocations	In-year reallocations	Medium-term capital budget realignment	Tax increases	Deficit financing	International assistance	Contingent credit	Insurance of public assets	Public support of private policy holder insurance	Parametric sovereign insurance	Catastrophic bonds
Brunei Darussalam											
Cambodia											
Indonesia											
Laos											
Malaysia											
Myanmar											
Philippines											
Singapore											
Thailand											
Vietnam											

Source GFDRR (2012a)

to *ex-ante* risk-financing, as Table 4.4 shows. As on example, most Association of Southeast Asian Nations (ASEAN) member countries are currently involved in or actively considering contingent risk-financing or sovereign insurance.

Yet, while processes and procedures are being implemented, budgeted amounts remain rather small and inadequate to tackle the increasing burden from disaster risk (GFDRR 2012a).

4.3.2 Implementing Innovative Risk-Financing Measures

Risk-financing through insurance and other hedging instruments spreads and pools risks, thus lessening the variability of losses, but it does not directly reduce risk. By providing indemnification in exchange for a premium payment, insured victims benefit from the contributions of the many others who are not affected, and thus in the case of a disaster they receive a contribution greater than their premium payment. However, over the long run, insured persons or governments can expect to pay significantly more than their (expected) losses. This is because of the costs of insurance transactions and the capital reserved by insurance companies for potential losses (or reinsurance), as well as the financial return required for absorbing the risks. The "load" can be significant, or as much as 500 % of the pure risk (expected losses) (Froot 2001). Still, people buy insurance, and justifiably so, because of their aversion to (large) losses—that is, their concern regarding the volatility of possible outcomes. Insurance and other risk transfer instruments are thus justified by the concept of risk aversion and it is because of aversion to large risks that people are willing to pay for insurance.

Traditional or parametric/index-based *insurance* provides indemnification against losses in exchange for a premium payment. It is the most common form of risk transfer, and there are well-established markets. The disadvantage is that the premium can be significant and is a definite cost against the budget.

In a *reserve fund*, amounts are laid aside on an annual basis, so that capital can accumulate. The fund accumulates in years without catastrophes and can be used in the case of an event to finance losses. However, for a vulnerable country facing events that might cost more than its entire annual GDP, this is not practical. Even for larger economies, the fund may not be able to accumulate sufficiently before the first disaster occurs, and it always needs to be replenished after it has been used. There is also a real danger that the fund will be "raided" for other purposes if a period without disasters creates a sense of false security.

Contingent credit arrangements do not transfer risk but spread it inter-temporally. In exchange for an annual fee, the right is obtained to take out a specific loan amount post-event that has fixed conditions. Contingent credit options are commonly grouped under alternative risk transfer instruments. The World Bank has recently developed such an instrument, now labelled a "deferred drawdown option". The disadvantage is that the exercise of the right creates a new debt, which can constrain future development.

Important innovations have also been implemented with respect to implementation disaster risk-financing in different regions.

The Ethiopian weather derivative: To supplement and partly replace the traditional food aid response to famine of the Ethiopian government as aided by the World Food Programme (WFP), WFP designed an index-based insurance system to provide extra capital in the case of extreme drought. The amount is based on contractually specified catastrophic shortfalls in precipitation measured in terms of the Ethiopia Drought Index (Wiseman and Hess 2007).

Mexico: FONDEN and the catastrophe bond: In 1996, the Mexican government created a budgetary programme called FONDEN (the Fund for Natural Disasters) to enhance the country's financial preparedness for natural disasters. FONDEN's objective is to prevent imbalances in federal government finances as a result of natural catastrophes. In 2006, the Mexican government chose to insure FONDEN against major earthquakes with a mix of reinsurance and a catastrophe bond, thus accessing both reinsurance and financial markets (Cárdenas et al. 2007)

CCRIF: The Caribbean island states in 2007 formed the world's first multi-country catastrophe insurance pool, reinsured in the capital markets, to provide governments with short-term liquidity in the aftermath of hurricanes or earthquakes. Sixteen Caribbean countries contribute resources ranging from $0.2 to $4 million depending on the exposure of their specific country to earthquakes and hurricanes. CCRIF has created a viable insurance instrument, and is helping improve the region's capacity to deal with disasters. Also, country risk profiles via the Multi-Peril Risk Evaluation System catastrophe risk modelling platform are under way, providing a systematic basis and entry point for more detailed information (GFDRR 2011).

4.4 Towards Comprehensive Disaster Risk Management

Approaches organised around the protection of the balance sheet using risk-financing instruments have seen a great deal of emphasis in disaster-prone countries. Yet can these lead into broadly supporting DRM? We discuss entry points and evidence.

4.4.1 Integrating Risk-Financing with Risk Reduction and Reconstruction

Figure 4.7 exhibits the different phases of disaster management, suggesting the various links from risk-financing to risk reduction as well as to preparedness and response and finally to dealing with "surprise". Today, DRM is still strongly focused on *ex-post* response, and spending on post-disaster recovery and reconstruction dwarfs the uptake of *ex-ante* risk management. The global information provided by

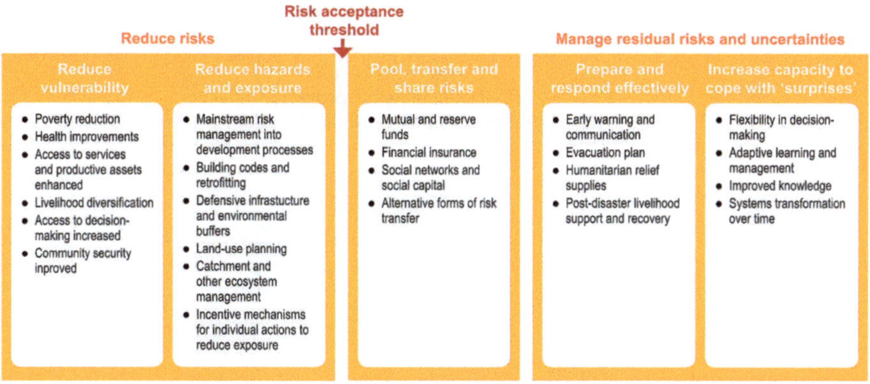

Fig. 4.7 Comprehensive DRM approach. *Source* Lal et al. (2012)

Kellett and Caravani (2013) of a ratio of 87–13 % in favour of *ex-post* response over *ex-ante* risk reduction is mirrored by case study information for this report for Mexico, with a balance of *ex-post* to *ex-ante* interventions of 90 versus 10 %.

Determining how much should be invested in risk reduction and how much in risk-financing as well as finding a proper balance between *ex-post* and *ex-ante* disaster management is not straightforward. It ultimately depends on the wider costs and benefits of both types of activities in terms of their interaction (e.g. financial instruments, through incentives, can influence prevention activities; see Linnerooth-Bayer et al. 2011) and their acceptability. Cost and benefits, in turn, depend on the nature of the hazard and risk. One way to think about the balance is illustrated by the risk-layering approach shown in Fig. 4.8.

For the low- to medium-loss events that happen relatively frequently, risk reduction is likely to be cost-effective in reducing burdens. The reason for this is that the costs of risk reduction often increase disproportionately with the severity of the consequences. Moreover, individuals and governments are generally better able to finance lower-consequence events (disasters) using their own means, for instance savings or calamity reserve funds, and including international assistance. The opposite is generally the case for risk-financing instruments, including reserve funds, catastrophe bonds and contingent credit arrangements. For this reason, it is generally advisable to use these instruments mainly for lower-probability hazards that have debilitating consequences (catastrophes). Finally, as shown in the uppermost layer of Fig. 4.8, individuals and governments will generally find it too costly to use risk-financing instruments against very extreme risks occurring less frequently than, say, every 500 years.

Budgetary policies and risk-financing options can in principle also lead to incentives for giving stronger emphasis to risk reduction. Implementing a structured process for risk detection in the balance sheet has the potential to provide a "price signal". In turn, a strong focus on *ex-post* disaster management (the still somewhat dominant approach, as discussed before) offers little in the way of risk awareness and stimulating the reduction of risk (Phaup and Kirschner 2010).

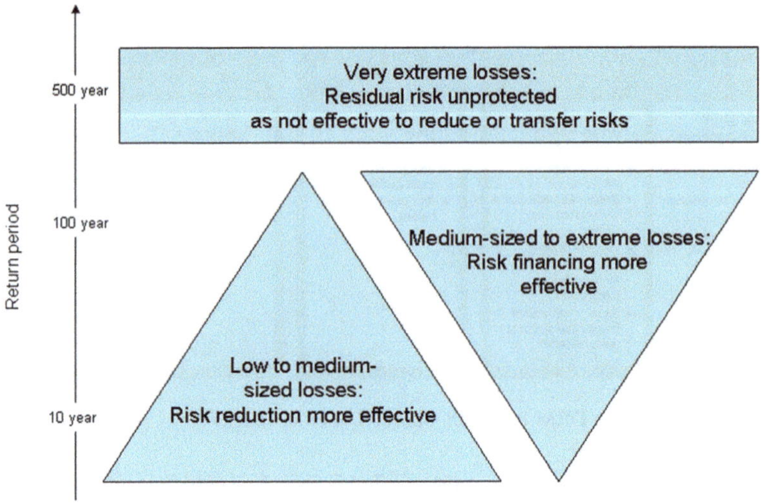

Fig. 4.8 The layering approach for risk reduction and risk-financing. *Source* Adapted from Mechler et al. (2014a)

While there is no detailed information on these linkages and incentives, some evidence from Mexico, one of the prime actors in fiscal risk management, provides for important learning. The primary interest in Mexico in DRM in the late 1990s was in identifying sovereign insurance to increase fiscal stability. In 1996, the government created its budgetary programme to enhance financial preparedness for natural disasters—FONDEN. Over the years, FONDEN has led to innovative risk financing arrangements, such as using catastrophe bonds to protect the balance sheet. As an ancillary benefit of the risk-financing strategy, which also required detailed information from risk assessments, risk reduction has been incentivised. FONDEN is currently promoting DRM in reconstruction activities, and about 25 % of its resources are earmarked to post-event rebuild damaged assets back better against future disasters. As another measure, with increasing risk and cost awareness, FONDEN is considering the relocation of housing in high-risk areas. Yet FONDEN's reach is limited because it is not a government agency but a financial instrument (personal communication; World Bank 2013).

4.4.2 Informing the Transition to Holistic Disaster Risk Management Integrated with Development: The Need for Broader-Based Decision-Making Tools

Moving from risk detection to risk-financing, there is stronger emphasis on comprehensive DRM. Where is the transition in thinking and implementation leading, and what tools can help support this shift in mind-set?

Debate regarding public sector DRM has largely focused on the use of economic efficiency-oriented approaches, which can be analysed using cost–benefit analysis (CBA). Over the years, appraisals of public investment decisions building on this logic have mushroomed and improved in terms of methodology. Recent analysis (Mechler et al. 2014b) highlights that CBA and associated processes can be very useful in supporting risk reduction decision-making, if key challenges are properly tackled. These challenges include complexities in estimating risk; data dependency of results; negative effects of interventions; inclusion of stakeholders; and distributional aspects. How this information is used will qualify the acceptability and robustness of the studies.

Key challenges remain and need attention, within these the consideration of intangibles, including multiple objectives such as equity and distributional issues, as well as taking a stronger systems perspective on the benefits, which means understanding how broad-based interventions in health, education and infrastructure can create cross-sectoral benefits. As Fig. 4.9, building on cost–benefit information on the returns of public interventions in various sectors, suggests, this is needed. The figure suggests investments *within* sectors such as health, nutrition, water and DRM all reap good returns well beyond the necessary condition of exceeding the benefit–cost threshold of 1. However, decision-makers, particularly in the finance ministry, faced with limited resources are left wondering how to create returns synergistically *across* sectors, which involves enhanced thinking about mainstreaming DRM into development and resilience-based strategies that can lead to co-benefits.

The need for further integration and mainstreaming of DRM into broader development agendas requires rethinking of the strategy and decision tools used to inform the strategy. For this, the use of a single "efficiency" criterion (as used by CBA) is becoming increasingly obsolete: more integrative decision-making frameworks that incorporate additional criteria, such as "co-benefits", "robustness" and "public acceptability", are increasingly needed. Such broader framing

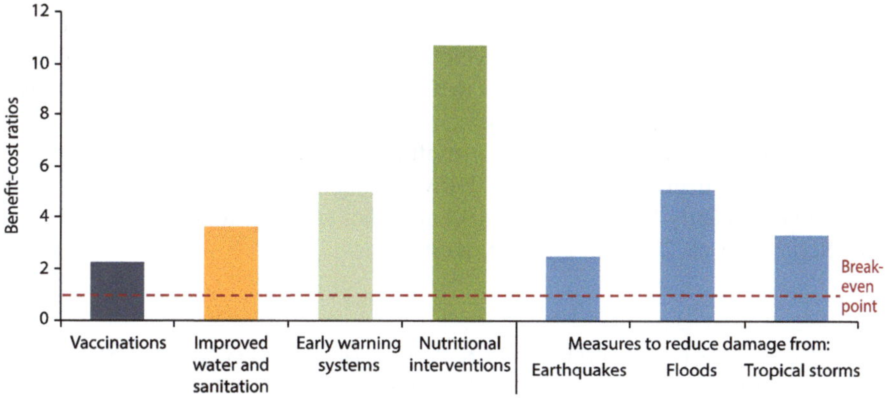

Fig. 4.9 Cost–benefit ratios of measures in various sectors. *Source* World Bank (2013)

Table 4.5 Characteristics and applicability of different decision support tools for assessing DRM

	Opportunities	Challenges	Typical application
CBA	Quantitative framework based on comparing costs with benefits under a single objective economic efficiency criterion	Need to monetise all benefits, difficulty in representing intangible impacts, such as value of life, value judgments of analyses not always fully transparent	Well-specified *hard resilience* projects with economic benefits
CEA	Ambition level fixed, and only costs to be compared. Intangible benefits, particularly loss of life, do not need to be monetised	Ambition level needs to be fixed and agreed on	Well-specified interventions with important intangible impacts, which should not be exceeded (loss of life, etc.)
MCA	Consideration of multiple objectives and plural values	Multiple criteria require weighting involving multiple value judgments, which can make replication complex	Multiple and systemic interventions involving plural values (e.g. investing in infrastructure and education)

Source Mechler et al. (2014a), Surminski (2014)

may colloquially be understood as a shift from risk to resilience thinking: policy-makers, practitioners and researchers are increasingly called to look beyond direct risk and to find critical linkages to the development–risk nexus. The IPCC recently identified an on-going shift in thinking with reference to climate change adaptation (where DRM figures prominently): economic analysis is moving away from a unique emphasis on efficiency, market solutions and CBA of adaptation to include consideration of non-monetary and non-market measures, risks, inequities and behavioural biases, and barriers and limits and consideration of ancillary benefits and costs (Chambwera et al. 2014).

This implies also looking beyond CBA to other tools available that can help public sector decision-makers to make decisions on DRM, such as cost-effectiveness analysis (CEA), which does not require the monetisation of intangibles, and multi-criteria analysis (MCA), which allows for multiple objectives to be assessed concurrently (Table 4.5).

MCA in particular appears to be a useful decision technique for the changing perspective on decision-making for DRM. While MCA thinking has not been applied significantly beyond frameworks and pilot studies,[2] it holds good potential (see Scrieciu et al. 2014 for a recent overview).

[2]MCA has been applied to DRM in the UN Environment Programme's project Multi-Criteria Analysis for Climate Change, commissioned to provide practical assistance to governments in preparing climate change mitigation and adaptation strategies. The objective was to assist government decision-makers, particularly in developing countries, to identify and examine policy

4.5 Towards Fiscal Resilience and Creating Co-benefits

Bridging gaps in integrating government risk-financing with risk reduction and with economic and development planning holds potential for putting a cost on risk and incentivising investments in risk reduction (Mitchell et al. 2014). However, stepping beyond a focus on DRM only, how can fiscal co-benefits be considered and created by following a synergistic strategy that focuses on both DRM and development? We discuss recent discourse on risk and resilience, then turn to entry points with relevance for the fiscal perspective.

4.5.1 A Broadening Discourse on Risk and Resilience

The DRM discourse is broadening framed around a resilience perspective. There is wide debate as to what such resilience framing will entail, but Keating et al. (2014: 8) suggest there is an emerging, if tacit, consensus that sees resilience as essentially forward looking in terms of "the ability of a system, community, or society to pursue its social, ecological, and economic development and growth objectives, while managing its disaster risk over time in a mutually reinforcing way" (see also UNESCAP 2013).

In the future, it is clear many countries will need to build their resilience to adapt and thrive in an unpredictable and shock-prone environment. To achieve this, they will need to make policy in a different way. Rather than dealing with problems in the economy, the environment and society separately, they will have to address these as parts of an overall system.

Similarly, in the climate change domain, the IPCC's Fifth Assessment Report Working Group II contributed to the reframing of climate change adaptation with regard to extreme climate events by emphasising risk management as fundamental to the policy response. The report suggests as the basis for policy action a shift towards the essentiality of managing extreme event risks holistically (to which climate change is contributing, in addition to other factors), rather than keeping a climate lens with a focus on climate adaptation policy only (Field et al. 2014).

Synergistic policy and pursuing co-benefits in programme and project planning may lead to impact in terms of increased investment in DRM. A recent evaluation by the Asian Development Bank (ADB) Independent Evaluation Department reports a significant number[3] of loans and grants disbursed by ADB for supporting DRM versus supporting disaster recovery over the time period 1995–2011 (ADB 2012).

(Footnote 2 continued)

options and measures for climate change that are low cost, environmentally effective and in line with national development priorities (http://www.mca4climate.info).

[3]Projects including dedicated DRM projects as well as other projects that incorporate and support building resilience.

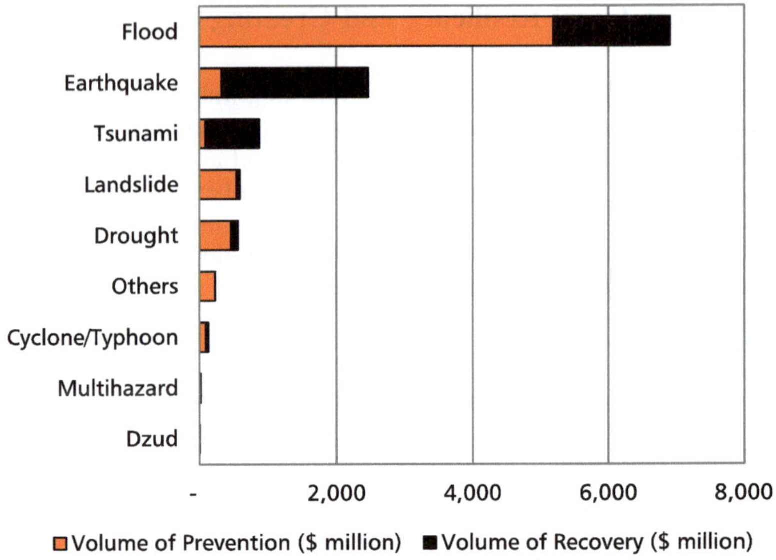

Fig. 4.10 Breakdown of ADB's DRM and recovery projects according to hazard-type. *Source* ADB (2012)

The breakdown reported between spending on disaster risk reduction projects versus disaster recovery has been 57–43 % in favour of financial support for pre-disaster activities, and 21 % predominantly allocated to risk management. What explains this surprisingly large share of DRM in disaster management shown in Fig. 4.10?

An important factor has been that, of the DRM-related projects, the majority of lending has been undertaken to partially or predominantly support flood risk management as part of water resource management, irrigation and drainage efforts. This integration seems to explain why the share of prevention versus recovery is a magnitude higher as compared with the global evidence on disaster spending. While lending occurred at substantially lower levels, landslide and drought DRM projects seemingly profited from a similar integrative strategy. For seismic and tsunami risk, co-benefits were perceived as small or less visible, and lending shows a strong reactive bias.

If a broader perspective is to be operationalised, what are the entry points for deliberative strategies for creating fiscal co-benefits? Based on the review of fiscal risk management approaches, two, not mutually exclusive, entry points emerge: fiscal disaster risk assessment leading to the mainstreaming of DRM; and broad-based contingency planning. Both, albeit with limited evidence, have the potential to lead to a broader co-benefit approach for dealing with disaster risk.

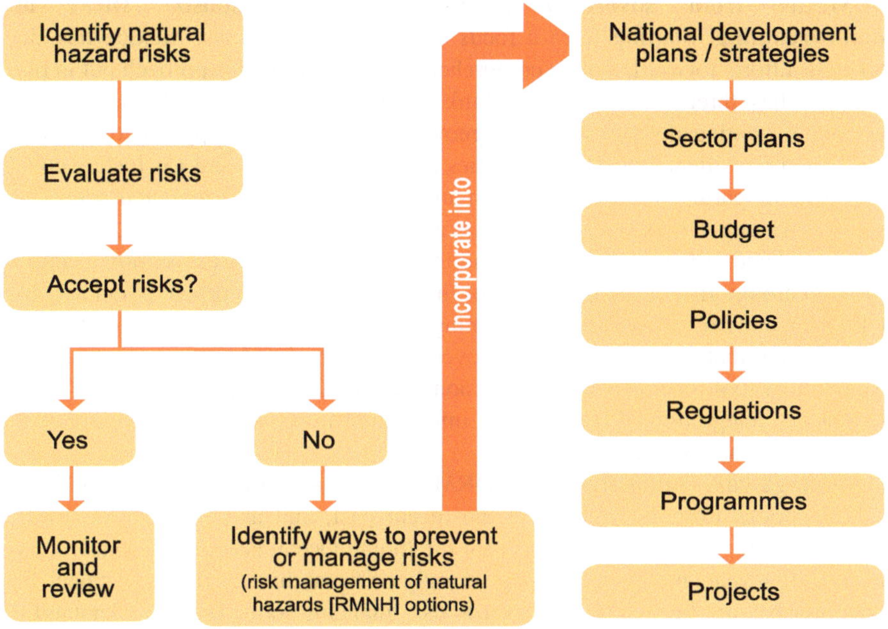

Fig. 4.11 Incorporating disaster risk assessments into strategies and plans. *Source* Bettencourt et al. (2006)

4.5.2 Disaster Risk as the Entry Point: Fiscal Disaster Risk Assessment and Mainstreaming

Mainstreaming DRM (and climate change adaptation) into development planning and policy has become the imperative, and features also in the Sendai Framework for Action (UN 2015).[4] Figure 4.11 identifies the rationale and process of mainstreaming risk in terms of factoring risk, if estimated to be important, into development-relevant planning at different levels, such as national programming and sectoral and budgetary planning. The budget process holds high appeal as it provides for links between national development and sectoral planning, and policies, regulations, programmes and ultimately projects.

We discuss approaches on mainstreaming that have foundations in fiscal disaster risk assessments.

[4]For example, "Promote the mainstreaming of disaster risk assessments into land-use policy development and implementation, including urban planning, land degradation assessments and informal and non-permanent housing, and the use of guidelines and follow-up tools informed by anticipated demographic and environmental changes" (15).

Mexico—from sovereign insurance towards holistic DRM and mainstreaming: Starting from a focus on insurance solutions, Mexico has since taken its efforts forward. A new comprehensive programmatic approach put in place in 2012 has three pillars (World Bank 2014). In addition to (1) strengthening Mexico's existing disaster risk management systems, it (2) supports joint disaster and climate resilience-building activities across key sectors and (3) fosters collaboration and partnership-building with many actors domestically and within the region. Among others, engagement occurs between the Ministry of Agriculture, Territorial and Urban Development and the National Disaster Prevention Centre on mainstreaming risk reduction policies into territorial and urban planning; with education authorities around strengthening safe school approaches; and on fostering partnerships that assess and tackle poverty with improved catastrophe risk management. A tool for information provision is a risk-modelling platform that aims at systematic integration of disaster risk information into the formulation and evaluation of federal investments.

Madagascar—mainstreaming DRM across sectors: Having experienced severe shocks from cyclones over the recent past, and as part of work towards setting up a regional DRM platform for the Indian Ocean islands, Madagascar over the past few years has been focusing strongly on fiscal disaster risk assessment. The intention has not been to work towards risk-financing tools, but to understand the budgetary implications of disaster risk and to identify options for managing these broadly. Given the importance of risk, the country has further mainstreamed risk into different sectors. Officially, the authority for DRM sits with the Prime

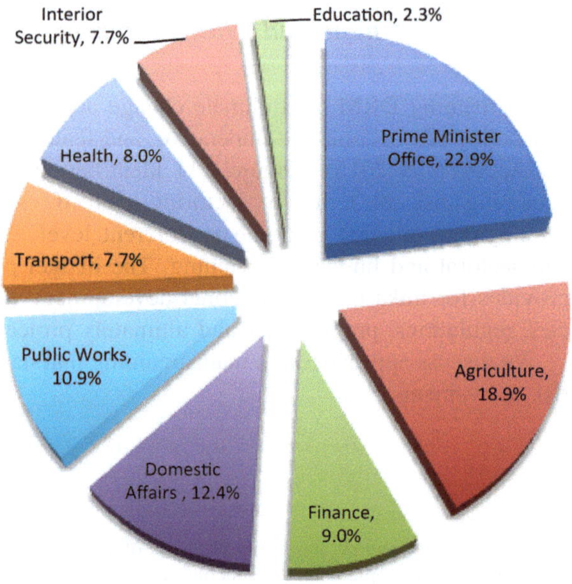

Fig. 4.12 DRM spending per ministry budgets (average 2010–2014). *Source* UNISDR (2015b)

Minister's Office, which, with the finance ministry, has been closely engaged on the budgetary risk analysis. Building on increasing risk awareness, DRM has increasingly become a crosscutting concern, and line ministries such as agriculture (a key risk is the loss of revenue following the physical loss of the vanilla export crop), domestic affairs, public works and transport are pursuing investment in DRM (see Fig. 4.12).

4.5.3 Holistic Fiscal Stress Testing and National Risk Assessments as Entry Points

Another, related, approach pursued with substantial effort is working towards a co-benefits approach via the fiscal risk matrix by considering many contingent risks and their interaction with disaster risk at the same time. Such a push has come from insights gained during the recent and on-going financial and fiscal crises. In the aftermaths, fiscal risks are being more systematically assessed, through sensitivity tests on baseline macro and fiscal indicators, commonly refereed to as stress testing. Also, there has been increased understanding of a need to take a systemic perspective to understand the potential for complex and interrelated shocks, essentially leading to a multi-risk approach (WEF 2015). Disaster risk has come to be considered a key threat, and in a recent survey regarding relevant fiscal risks in OECD countries disasters came out as an important concern (Table 4.6).

Colombia and the UK are examples of countries that have started to pursue broader multi-risk strategies in fiscal and public risk management.

Colombia—towards broad-based fiscal risk management: Fiscal risk assessment has become an important consideration for working towards a more sustainable and equitable development strategy. Fiscal risk assessment has become mandatory in Colombia and disaster risk, ranked the second most relevant risk, is seen as a critical component of a broader fiscal risk management strategy, which looks at the various risks that are interlinked and options for mutually managing

Table 4.6 Relevance of disaster risk for fiscal management in OECD countries

Category	Relevance			
	High	Medium	Low	None
	In % of each category			
Pension funds	6	37	31	25
Public–private partnerships and other risk-sharing	6	16	44	34
Financial sector	31	22	25	22
Legal claims	3	9	53	34
Other liabilities and guarantees	9	19	63	9
Natural disasters, health care risks	9	13	53	25

Source Kopits (2014)

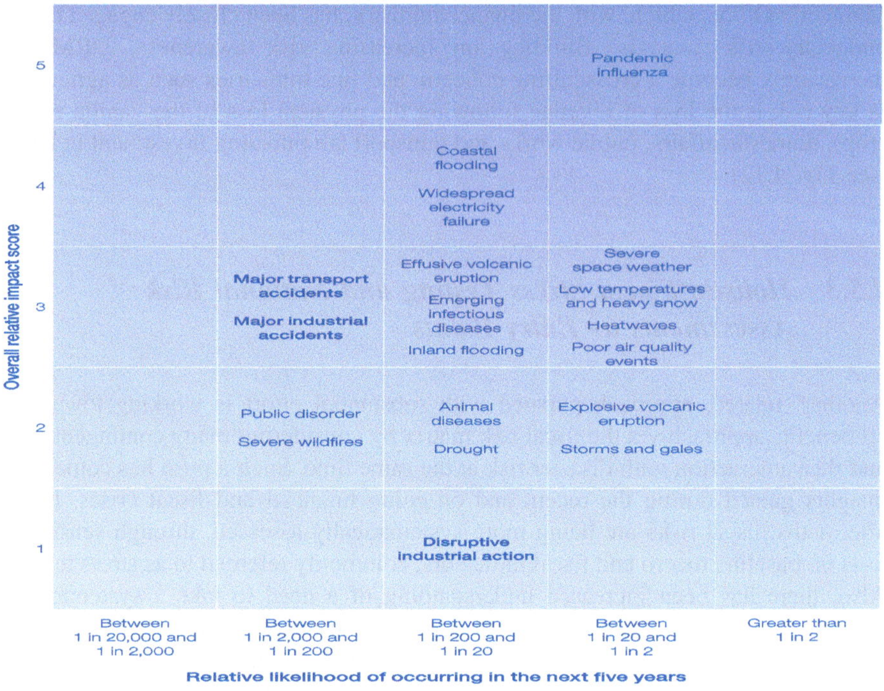

Fig. 4.13 UK's risk matrix for 2015. *Note* Terrorist risks are visualised separately in the report. *Source* UK Cabinet Office (2015)

risk across issues of concern. As one example, the government of Colombia is intent on upgrading catastrophe insurance requirements for concessions. This would help reduce its contingent liabilities that arise from public–private partnership arrangements undertaken for infrastructure construction and operation (World Bank 2011).

UK—national risk assessments as broad-based planning tools for multi-risk strategies: The UK since 2008 (and similarly the Netherlands since 2007) has taken a broad-based perspective on risks throughout. National risk assessments to improve policy related to preventing and planning for key risks (such as health-related or terrorist-focused risks) have been undertaken bi-annually by the UK Cabinet Office since 2008 and are being published as National Risk Registers (UK Cabinet Office 2015). These assessments identify and measure main risks bearing on the country—natural, technological, terrorist and other types of risk, following a systematic methodology of risk identification, scenario-building and determination of impacts. The quantitative part is finally summarised by a national risk matrix (see Fig. 4.13), which organises the main risks according to probability of occurrence and impact. The synoptic representation of risks provides for a level playing field, which allows for planning policy measures and, in theory, linking of agendas, such as with the national climate change risk assessment, which has to be undertaken every five years as decreed by the Climate Change Act (see DEFRA 2012). However, it is currently not clear whether this analysis has truly led to the implementation of options (also, financial risks are not considered).

A key benefit of this comprehensive risk assessment exercise is seen in better allowing for coordination and cooperation as well as allocation across ministries and public sector organisations. Furthermore, such planning helps provide incentives for better managing risk *ex-ante*, as it anticipates the *ex-post* consequences and trade-offs involved in responding to shocks. As a case in point, in the Netherlands, cross-regional competition for resources for emergency management after large floods is a concern that has been recognised using the risk assessment. Finally, another key point, particularly for resilience-based strategies, is that such broad risk assessments allow the identification of new actors, including the private sector, which takes part in the assessments as well. Whether and how these comprehensive risk assessments are replicable in other places and regions with more limited capacity and resources remains an open question. Yet the government of Morocco with support from the World Bank and the Global Facility for Disaster Reduction and Recovery (GFDRR) has started undertaking a multi-risk exercise focused on natural disaster risk, commodity price shocks and agricultural sector risks. It is planned that this will lead into identifying options and considering relevant institutions to further implement this agenda (World Bank 2013).

4.6 Conclusions: From Fiscal Risk to Building Resilience to Harnessing Co-benefits

A large part of disaster risk ends up with the fiscal position, and there has been increasing recognition of the need to plan for disaster. Yet fiscal risk management is not an easy proposition, as disaster risk is a contingent liability—that is, costs accrue only in the case of an event. Furthermore, a large proportion of liabilities is of an implicit, unwritten nature (disaster relief and recovery to households and business), unlike with direct liabilities (reconstruction of lost infrastructure and assets).

Over the past few years, fiscal policy on and public investment in DRM in many countries exposed to disaster risk has seen a step change. Based on experiencing and better understanding the large fiscal and economic burdens of disasters, fiscal and development planning has graduated from a perspective of risk ignorance to one of risk awareness. This effectively means that, increasingly, risk is explicitly taken into account in fiscal decisions and is being considered part of contingency liability planning, indicating a shift in perspectives from a risk-neutral to a risk-averse planning stance.

Progress in fiscal risk planning has been achieved based on tools available to systematically assess and manage risks in the fiscal balance sheet (fiscal risk and hedge matrices). Better risk planning may lead to improved risk detection across sectors. Countries have started to develop broad risk matrices that chart out probability versus impact for many diverse risks, which helps in considering measures that broadly enhance fiscal stability. Reduced budgetary uncertainty allows governments to focus less on crisis management and more on longer-term issues.

At the same time, identifying fiscal risks *vis-à-vis* fiscal hedging instruments helps develop a level playing field for investments in DRM and other priority areas. Such systematic thinking has mostly informed considerations of sovereign insurance across highly exposed developing countries. Yet insurance is only one element in the DRM toolbox. It is widely recognised that, in the face of increasing risk, a broad-based perspective is necessary to incentivise risk reduction, avoid risk creation and generate co-benefits that go beyond direct and indirect gains by creating a third dividend "beyond disasters", that contributes to providing resilience against shocks more holistically.

This discussion has traced the development of fiscal DRM around four steps. These steps and activities may lead to three dividends, as framed in the project overall, as follows:

1. Understanding fiscal risk
2. Protecting public finance through risk-financing instruments (first dividend)
3. Working towards comprehensively managing disaster risk, including risk reduction and risk preparedness as they affect development (second dividend)
4. Pursuing a synergistic strategy of managing disaster risks and promoting development (third dividend).

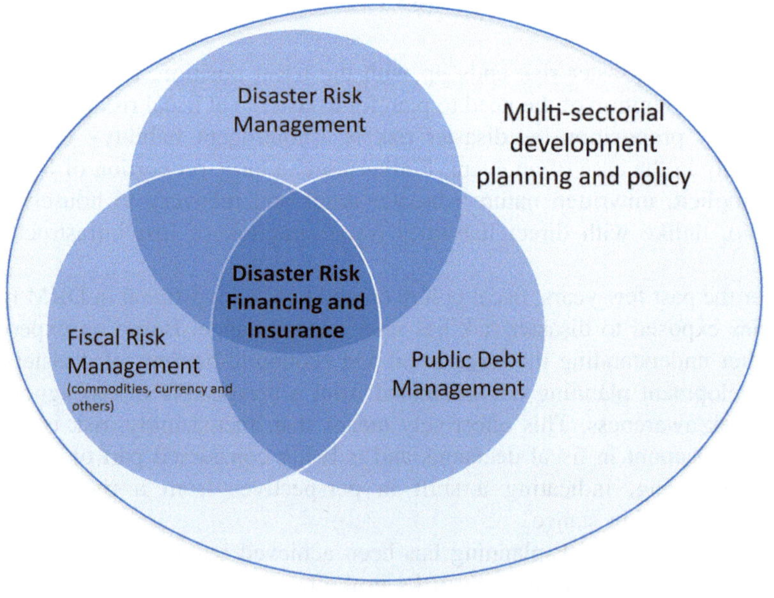

Fig. 4.14 Suggested integration of DRM with fiscal risk management, public debt management and development policy and planning. *Source* Adapted from Holm-Nielsen (2012)

Table 4.7 Adopted strategy and entry points for synergistic co-benefits strategies

Country	Entry point	Strategy
Madagascar	Fiscal disaster risk assessment	Fiscal disaster risk assessment leading to mainstreaming DRM
Colombia	Fiscal disaster risk assessment and sovereign risk financing	From fiscal disaster risk assessment towards a broad fiscal risk management strategy
Mexico	Sovereign disaster risk financing	Sovereign insurance leading into comprehensive DRM and mainstreaming
UK, Netherlands (Morocco)	National risk assessment	Multi-risk planning for synergistic risk-based policies

As we have shown, all steps and foci are seeing some activity: Steps 1 and 2 have been implemented in a number of countries, and Step 3 is increasingly being tackled, whereas Step 4 will need more attention in the future to truly create measureable co-benefits and build resilience throughout. There is increasing recognition that a broad-based perspective is necessary to incentivise risk reduction, avoid risk creation and generate additional co-benefits that go beyond the direct and indirect gains from reducing risk. Co-benefits can be achieved through better integration of DRM with fiscal risk management, public debt management and development policy and planning, as Fig. 4.14 suggests.

Our discussion tentatively suggests that fiscal disaster stress testing and national risk assessment can be entry points for more holistically tackling DRM and development in terms of a co-benefits strategy. These two entry points were found to be active for a limited number of countries, as summarised in Table 4.7.

The potential co-benefits of fiscal DRM overall would comprise, among others,

- Improved planning processes for contingencies, providing the grounds for synergistic investments into various sectors at the same time
- Solid returns from managing multiple stresses and shocks at reduced cost. For example, sorely needed investments into health and infrastructure often help build disaster resilience; at the same time, mainstreaming disaster risk reduction into these sectoral investments helps safeguard any benefits that will accrue despite strong exposure to shocks
- Risk planning helps with improving risk detection across sectors and identifying key public and private sector actors for managing risks. As one example, the UK has developed a risk matrix that charts out probability versus impact for many diverse risks, and thus makes it possible to take options that cut across sectors and broadly enhance resilience

Analytical insight and tools are key to supporting this on-going transition, and we identified fiscal stress testing, national risk assessments and multi-metric evaluations as important elements of a broad-based toolbox, which, if applied with sufficient stakeholder involvement across local to national to international scales, can help in working towards creating strong dividends that help us better manage disaster risk.

References

ADB (Asian Development Bank) (2012) ADB's response to natural disasters and disaster risks. Special Evaluation Study. ADB, Manila

Anginer D, de la Torre A, Ize A (2013) Risk-bearing by the state: when is it good public policy? J Financ Stab 10:76–86

Arrow KJ, Lind RC (1970) Uncertainty and the evaluation of public investment decisions. Am Econ Rev 60:364–378

Bettencourt S, Croad R, Freeman P, Hay J, Jones R, King P, Lal P, Mearns A, Miller G, Pswarayi-Riddihough I, Simpson A, Teuatabo N, Trotz U, Van Aalst M (2006) Not if but when-adapting to natural hazards in the Pacific Islands region: a policy note. East Asia and Pacific Region. Pacific Islands Country Management Unit. The World Bank, Washington D.C.

Cárdenas V, Hochrainer S, Mechler R, Pflug G, Linnerooth-Bayer J (2007) Sovereign financial disaster risk management: the case of Mexico. Environ Hazards 7:40–53

Chambwera M, Heal G, Dubeux C, Hallegatte S, Leclerc L, Markandya A, McCarl BA, Mechler R, Neumann JE (2014) Economics of adaptation. In: Field C, Barros V, Dokken DJ, Mach KJ, Mastrandrea MD, Bilir E, Chatterjee M, Ebi K, Estrada Y, Genova RC, Girma B, Kissel ES, Levy AN, MacCracken S, Mastrandrea PR, White LL (eds) Climate change 2014: impacts, adaptation, and vulnerability. Part A: Global and sectoral aspects. Cambridge University Press, Cambridge and New York

Cummins JD, Mahul O (2008) Catastrophe risk financing in developing countries: principles for public intervention—overview. World Bank, Washington, DC

DEFRA (UK Department for Environment, Food & Rural Affairs) (2012) UK climate change risk assessment 2012. DEFRA, London

Field C, Barros V, Dokken DJ, Mach KJ, Mastrandrea MD, Bilir E, Chatterjee M, Ebi K, Estrada Y, Genova RC, Girma B, Kissel ES, Levy AN, MacCracken S, Mastrandrea PR, White LL (eds) (2014) Summary for Policymakers. In: Climate change 2014: impacts, adaptation, and vulnerability. Part A: Global and sectoral aspects. Contribution of Working Group II to the Fifth Assessment Report (AR5) of the IPCC. Cambridge University Press, Cambridge and New York

Field C, Barros V, Stocker T, Dahe Q, Dokken DJ, Ebi KL, Mastrandrea MD, Mach KJ, Plattner G-K, Allen SK, Tignor M, Midgley PM (eds) (2012) Summary for policy makers. In: Special report on managing the risks of extreme events and disasters to advance climate change adaptation: Summary for Policymakers. Report of Working Groups I and II of the IPCC. Cambridge University Press, Cambridge and New York

Froot K (2001) The market for catastrophe risk: a clinical examination. J Financ Econ 60(23):529–571

GFDRR (Global Facility for Disaster Reduction and Recovery) (2011) Caribbean Catastrophe Risk Insurance Facility (CCRIF). GFDRR, Washington, DC

GFDRR (Global Facility for Disaster Reduction and Recovery) (2012a) Advancing Risk Financing and Insurance in ASEAN Member States: Framework and Options for Implementation, vol 1. World Bank, Washington, DC

GFDRR (Global Facility for Disaster Reduction and Recovery) (2012b) Fiscal Management of Natural Disasters in Colombia. GFDRR, Washington, DC

Ghesquiere F, Mahul O (2007) Sovereign natural disaster insurance for developing countries: a paradigm shift in catastrophe risk financing. Policy Research Working Paper 3278. World Bank, Washington, DC

Gurenko E (2004) Introduction. In: Gurenko E (ed) Catastrophe risk and reinsurance: a country risk management perspective. Risk Books, London

Hochrainer-Stigler S (2006) Macroeconomic risk management against natural disasters. DUV, Wiesbaden

Hochrainer-Stigler S, Mechler R, Pflug G, Williges K (2014) Funding public adaptation to climate-related disasters. Estimates for a global climate fund. Glob Environ Change 25:87–96

Holm-Nielsen N (2012) Fiscal Risks Related to Catastrophes in the Caribbean. Presentation

Hourcade J-C, Shukla P (2001) Global, regional, and national costs and ancillary benefits of mitigation. In: Contribution of working group III to the third assessment report of the IPPC. Cambridge University Press, Cambridge and New York

IADB (Inter-American Development Bank) (2008) The disaster deficit index (DDI). IADB, Washington, DC

Keating A, Campbell K, Mechler R, Michel-Kerjan E, Mochizuki J, Kunreuther H, Bayer J, Hanger S, McCallum I, See L, Williges K, Atreya A, Botzen W, Collier B, Czajkowski J, Hochrainer S, Egan C (2014) Operationalizing resilience against natural disaster risk: opportunities, barriers, and a way forward. White Paper. Zürich Flood Resilience Alliance, Zürich

Kellett J, Caravani A (2013) Financing disaster risk reduction: a 20-year story of international aid. ODI and GFDRR, World Bank, London and Washington, DC

Kopits G (2014) Coping with fiscal risk: analysis and practice. OECD J Budgeting 14(1):47–71

Lal P, Mitchell T, Aldunce P, Auld H, Mechler R, Miyan A, Romano LE, Zakaria S (2012) National systems for managing the risks from climate extremes and disasters. In: Field C, Barros V, Stocker T, Dahe Q, Dokken DJ, Ebi KL, Mastrandrea MD, Mach KJ, Plattner G-K, Allen SK, Tignor M, Midgley PM (eds) Managing the risks of extreme events and disasters to advance climate change adaptation. Intergovernmental Panel on Climate Change Special Report. Cambridge University Press, Cambridge and New York

Linnerooth-Bayer J, Hochrainer S, Mechler R (2011) Insurance against Losses from Natural Disasters in Developing Countries. Evidence, Gaps and the Way Forward. J Integr Disaster Risk Manag. doi:10.5595/idrim.2011.0013

Linnerooth-Bayer J, Mechler R, Pflug G (2005) Refocusing disaster aid. Science 309:1044–1046

Mechler R (2004) Natural Disaster Risk Management and Financing Disaster Losses in Developing Countries. Risikoforschung und Versicherungsmanagement, Karlsruhe

Mechler R, Bouwer L, Linnerooth-Bayer J, Hochrainer-Stigler S, Aerts J, Surminski S (2014a) Managing Unnatural Disaster Risk from Climate Extremes. Nature Climate Change 4:235–237

Mechler R, Czajkowski J, Kunreuther H, Michel-Kerjan E, Botzen W, Keating A, McQuistan C, Cooper N, O'Donnell I (2014b) Making communities more flood resilient: the role of cost benefit analysis and other decision-support tools in disaster risk reduction. White Paper. Zürich Flood Resilience Alliance, Zürich. http://opim.wharton.upenn.edu/risk/library/ZAlliance-decisiontools-WP.pdf

Mechler R, Hochrainer-Stigler S (2014) Revisiting Arrow-Lind: managing sovereign disaster risk. J Nat Resour Policy Res 6(1):93–100

Mitchell T, Mechler R, Harris K (2014) Disaster risk management and adaptation to extreme events: placing disaster risk management at the heart of national economic and fiscal policy. In: Markandya A, Galarraga I, de Murieta ES (eds) Handbook on the economics of adaptation. Routledge, New York

Musgrave RA (1959) The theory of public finance. McGraw Hill, New York

Noy I (2009) The macroeconomic consequences of disasters. J Dev Econ 88(2):221–231

OECD (Organisation for Economic Co-operation and Development) (2012) Disaster risk assessment and risk financing. G20/OECD Methodological Framework. OECD, Paris

Phaup M, Kirschner C (2010) Budgeting for disasters: focusing on the good times. OECD J Budgeting 1:1–24

Polackova Brixi H, Mody A (2002) Dealing with government fiscal risk: an overview. In: Polackova Brixi H, Mody A (eds) Government at risk. World Bank, Washington, DC

Priest GL (2003) Government insurance versus market insurance. Geneva Papers Risk Insur 28:71–80
Scrieciu Ş, Silviu ZC, Belton V, Mechler R (2014) Advancing methodological thinking and practice for development-compatible climate policy planning. Mitig Adapt Strat Glob Change. doi:10.1007/s11027-013-9538-z
Surminski S (2014) Decision-Making tools. Concept Note. GFDRR and ODI, Washington, DC, and London
Tanner T, Surminski S, Wilkinson E, Reid R, Rentschler J, Rajput S (2015) The Triple Dividend of Resilience: Realising development goals through the multiple benefits of disaster risk management. Global Facility for Disaster Reduction and Recovery (GFDRR) at the World Bank and Overseas Development Institute (ODI), London
UK Cabinet Office (2015) National risk register of civil emergencies, 2015th edn. UK Cabinet Office, London
UK GSO (UK Government Office for Science) (2012) Foresight. Reducing risks of future disasters: Priorities for decision makers. UK GSO, London
UN (2015) Sendai framework for disaster risk reduction 2015–2030. A/CONF.224/CRP.1
UNESCAP (UN Economic and Social Commission for Asia and the Pacific) (2013) Building resilience to natural disaster and major economic crises. UNESCAP, Bangkok
UNISDR (UN Office for Disaster Risk Reduction) (2013) Global Assessment Report on Disaster Risk Reduction 2013. UNISDR, Geneva
UNISDR (UN Office for Disaster Risk Reduction) (2015a) Global Assessment Report on Disaster Risk Reduction 2015. UNISDR, Geneva
UNISDR (UN Office for Disaster Risk Reduction) (2015b) Review of Madagascar. Working paper on public investment planning and financing strategies for disaster risk reduction. UNISDR, Geneva
WEF (World Economic Forum) (2015) The global risks report 2015. WEF, Davos
Wilkinson E (2012) Transforming disaster risk management: a political economy approach. ODI, London
Williges K, Hochrainer-Stigler S, Mochizuki J, Mechler R (2015) Modeling the indirect and fiscal risks from natural disasters: emphasizing resilience and 'Building Back Better.' Background Paper for the 2015 Global Assessment Report on Disaster Risk Reduction. UNISDR, Geneva
Wiseman W, Hess U (2007) Reforming humanitarian finance in Ethiopia: a model for integrated risk financing. Working Paper. WFP, Rome
World Bank (2010a) Natural hazards and unnatural disasters. UN and World Bank, Washington, DC
World Bank (2010b) Financial protection of the state against natural disasters. Policy Research Working Paper 5429. World Bank, Washington, DC
World Bank (2011) Analysis of disaster risk management in Colombia. A contribution to the creation of public policies. GFDRR, Washington, DC
World Bank (2013) World Development Report 2014. Risk and Opportunity. Managing Risk for Development. World Bank, Washington, DC
World Bank (2014) A novel approach to disaster risk management: the story of Mexico. World Bank, Washington, DC

Chapter 5
Capturing the Co-benefits of Disaster Risk Management in the Private Sector

Adam Rose

Abstract In most countries, the private sector owns the vast majority of the buildings and infrastructure at risk. However, most investment in disaster risk management (DRM) is made by the public sector, while the private sector lags far behind. This situation represents missed opportunities for businesses to capture not only higher levels of direct benefits of DRM but also a broader set of co-benefits for themselves and for society as a whole. These co-benefits include ways of lowering production costs, improving the health of workers and contributing to general economic stability. Ironically, many of them are more tangible and immediate than ordinary DRM benefits, which may not appear until a disaster has struck many years after the investment has been made. This chapter analyses several important facets of private sector investment in DRM, primarily from an economic perspective. It is intended as a first step towards promoting greater investment in DRM by identifying potential co-benefits, explaining why they are not always pursued and suggesting ways to integrate them into private sector decision-making. The latter include government incentives, justified on the grounds that many private sector investments have extensive co-benefits, many of which pay dividends to society as a whole.

Keywords Disaster risk management · Private sector · Market failure · Co-benefits · Spillover effects

---

The author wishes to acknowledge the helpful suggestions of Swenja Surminski, Philip Schneider, Adrien Vogt-Schilb and Marianne Fay during earlier drafts of this chapter.

A. Rose (✉)
Price School of Public Policy and Center for Risk and Economic Analysis
of Terrorism Events, University of Southern California, Los Angeles, CA, USA
e-mail: adam.rose@usc.edu

S. Surminski and T. Tanner (eds.), *Realising the 'Triple Dividend of Resilience'*, Climate Risk Management, Policy and Governance,
DOI 10.1007/978-3-319-40694-7_5

5.1 Introduction

In nearly every country of the world, the private sector comprises most of the economy. As such, it represents the major source of potential losses from disasters, and hence is the sector with the most at stake. In most countries, however, the government carries out the majority of disaster risk management. This chapter explains this mismatch and offers insights into how greater awareness of the triple resilience dividend can stimulate an increase in much-needed additional investment in the private as well as the public sector. This investment is beneficial not only to businesses but also to their host economies and the people they sustain. It has positive feedback effects as well, not just for the firm undertaking the investment but also for the private sector in general.

There is no doubt that the primary motivation of businesses is to maximise profits. On the surface, only those DRM actions that are viewed as promoting this objective will be undertaken, although not all of these actions are recognised because of a lack of information or split incentives in large firms, as we discuss below. But several broader benefits of DRM, for businesses themselves and for society and the economy as a whole, are typically neglected—a problem that contributes to continued underinvestment in DRM.

We refer to the broader benefits, or spillover effects, of DRM as "co-benefits". We use the term broadly in this chapter to encompass both the first and second dividends of the triple dividend (Tanner et al. 2015).[1] For example, when a business installs a sprinkler system to protect against fire spreading on its premises, it also helps protect adjacent buildings and an entire community. This is also the case for strengthening the foundation of its headquarters building, lest it collapse on its neighbours, or for instituting better water drainage around its facilities that reduces flooding potential for the community and hence makes the operation of its other businesses less risky. More broadly, actions by any one business can contribute to overall community wellbeing in terms of improving quality of life and promoting economic stability. The firm does not capture these co-benefits, and this leads to underinvestment in DRM. In this chapter, we explore these co-benefits and identify some guidelines and incentives government can implement to encourage private firms to increase investment in DRM.

Another source of co-benefits pertains to broader gains to society outside of disaster situations, in the context of the second dividend. This refers to DRM investments that benefit disadvantaged segments of the population and that contribute to sustainable development itself, such as businesses vaccinating their employees or using more durable materials in their buildings. Many of these gains are external to the firm but feed back positively onto businesses themselves by promoting economic stability, spawning a healthier and more productive workforce, etc.

[1]The reader will recall from pervious chapters that the three dividends of DRM are avoided disaster losses, unlocking economic potential through reduced risk and generating co-benefits (often characterised as "joint products").

Still another category of co-benefits can be captured primarily by businesses themselves, if they are able to recognise them. These include DRM investments that improve the image of the firm or that otherwise lead to an increase in long-run profits, such as actions that are viewed as being in the public interest or that protect society from catastrophic risk.[2] Ironically, many of these co-benefits are more tangible and immediate than most ordinary DRM benefits, which may not appear until a disaster has struck many years after the investment has been made.

This chapter analyses many important facets of private sector investment in DRM, primarily from an economic perspective. It is intended as a first step towards greater investment in DRM by identifying potential co-benefits, explaining why they are not always pursued and suggesting ways to integrate them into private sector decision-making.

5.2 Shortfalls in Private Sector Investment in Disaster Risk Management

5.2.1 The Components of Private Sector DRM

The first line of defence against a disaster is to minimise the probability that it will take place, followed closely by minimising losses should it happen. Many natural disasters, such as earthquakes, tornadoes and hurricanes, cannot be stopped; others, such as terrorist attacks or major technological failures, can be. Other hazards, such as floods, can in some cases be stymied in advance by proactive human actions, ranging from adequate sewage system capacity to land-use planning.

Losses from disasters are not confined to property damage. Moreover, property (the capital *stock*) is not valuable just in and of itself but also in terms of the *flow* of goods and services it generates, often measured in terms of gross domestic product (GDP). Disruption of this flow, or business interruption (BI), begins at the point of the disaster and continues until the economy has recovered (Rose 2009a). In several recent disasters, such as the 11 September 2001 World Trade Center bombings and Hurricane Katrina, BI losses have exceeded property damage (Rose et al. 2009).

Post-disaster resilience, in contrast with (pre-disaster) mitigation, can be considered a second line of defence. While many have defined resilience so broadly as to include all activities that reduce disaster losses, we confine our use of the term to those activities that dampen BI losses by using remaining resources more efficiently and recovering more quickly (Rose 2009a). Post-disaster resilience is capable of reducing ordinary losses; it cannot compensate for irreversibility, such as loss of human life and the destruction of unique ecosystems.

---

[2]These might be considered part of the second dividend, or as extended private benefits, rather than as spillover effects, which characterise both the second and the third dividends.

A further response mechanisms for the private sector is *adaptation*, a term we use to describe reactions to climate change manifestations that do occur, although we acknowledge the wide divergence in definitions in the literature (Field et al. 2012; IPCC 2014). It is frequently applied to the longer-term impacts (stressors), and thus can be thought of as the long-run counterpart to resilience, which is typically, but not always, applied to natural disasters (which in the atmospheric realm are related to short-run climate variability, or more generally shocks). This chapter confines its attention to DRM in response to disasters as a manifestation of short-term climate shocks and thus does not focus separately on long-term stressors. However, several of the analytical principles and policy remedies overlap. For example, Kull et al. (2013) point to index-based insurance, which requires identification of technical features of risk and hence technology to evaluate them, thereby improving risk assessments of both disasters and long-term manifestations of climate change.

Several analysts have suggested, however, that it is prudent for businesses to integrate DRM into their general business strategies and practices. Essentially, this is a general way of capturing some co-benefits. For example, Zolli and Healy (2012) emphasise the importance of organisational and operational flexibility, which makes businesses more capable of coping with drastic change. Sheffi (2005) provides numerous examples of supply-chain rigidity that have led to business failure in the aftermath of a disaster, but this view is now countered by a broader view of the supply chain as a web that includes back-up sources.

5.2.2 Private Sector Investment Decisions

The primary objective of businesses is to maximise profits, and for many firms the focus is only on the short term. Secondary objectives, such as increasing market share, are usually consistent with this primary concern but are longer in term. Businesses will undertake DRM as long as it is consistent with profit maximisation. Most DRM initiatives involve investment, so the objective is often couched in terms of maximising net present value or the internal rate of return. The latter principle is often used to compare or rank alternatives.

Investments have two interrelated features that other business activities do not have, or have to a lesser degree. First, the returns to the firm take place over the course of time. Second, because they take place in the future, they involve a degree of uncertainty. Interest rates are used to account for both features. Revenues in the future year are not directly comparable, so the market interest rate is used to discount future returns to account for the time value of money. The interest rate is also used to adjust for risk, such that higher-than-market rates reflect a *risk premium*. Investments in new or unproven technologies are relatively more uncertain than others; these are characteristic of many of the larger DRM investments, which have to be customised. Another factor that translates into greater uncertainty is the benefits of DRM. These are the avoidance of lost profits, adjusted by the probability of occurrence of the disaster. These probabilities are highly uncertain, so the risk premium may be increased accordingly.

A major debate revolves around whether uncertainty reduces or merely postpones business investment. Doh and Pearce (2004) suggests a context of continuous versus discontinuous uncertainty makes a difference, with disasters clearly being in the latter category. A great deal of literature relates to regulatory uncertainty, which is also very applicable to disaster mitigation (e.g. building codes, zoning ordinances). Findings are mixed, however, on whether uncertainty inhibits forward movement (Aragón-Correa and Sharma 2003; Yang et al. 2004). Carrera et al. (2003) and others note ways of capitalising on uncertainty, such as the advantages of being a "first mover" (another type of co-benefit). Hoffmann et al. (2009) provide an example of three factors in a case study of German electric utilities facing climate change regulation as influencing forward progress on investment: (1) securing competitive resources, (2) leveraging complementary resources and (3) alleviating institutional pressure.

5.2.3 Private Sector Investment Decisions in Relation to the Public Sector

Although businesses continuously voice their support of free markets and opposition to government interference, most firms will work in a cooperative spirit if it is in their best interest. For enlightened business managers, best interest means not just short-run maximisation of profits, but also maintenance of the business' image. Both of these factors augur well for businesses cooperating in sustainable development efforts.

Governments are increasingly integrating or "mainstreaming" DRM into development planning (Mitchell et al. 2010). Public sector investment decisions are not at the core dissimilar to those of the private sector. Cost–benefit analysis is typically applied with the objective of maximising the net present value of future returns. Given the uncertain nature of disasters, the benefit side of the ledger is often weighted by the probability of occurrence. In such an expected value setting, co-benefits for extreme events can readily exceed the direct benefits of DRM because the weights are so low (the events are so infrequent). It should be kept in mind, however, that, with regard to other decision criteria, such as minimising regret, the probabilities do not come into play, and avoidance of disaster is the key factor.

5.2.4 Private Sector Co-benefits

The major benefits to a firm from a DRM investment are the revenue losses avoided (first dividend) and any ancillary revenues it may receive (second and third dividends in addition to broader benefits to other businesses, households and society as a whole). The broader aspects of the third dividend can be exemplified by co-benefits in the form of saleable by-products or joint products, such as a decision to install solar panels to insulate the firm against disruptions from central power stations, where excess solar electricity can be sold back to the grid. This is still a relatively short-sighted perspective, and the following subsections discuss its limitations from the vantage point of both the firm and society as a whole.

A more enlightened view of the firm's objectives incorporates longer-run considerations relating to its good name or its survival. DRM often provides broader social benefits that the firm might consider, as pressures from shareholders, incentive systems, etc. tend to focus private companies on the bottom line. Examples are voluntary reductions in pollution that improve the firm's image and implementation of flood control practices that benefit the entire floodplain.

An in-depth discussion of distinctions between categories of co-benefits is presented elsewhere in this volume, so we confine ourselves to a brief summary list of the categorisations this chapter uses. We deem it worthwhile to extend the number of DRM categories to five to encompass two categories that economists often emphasise and estimate separately but that are only implicit in the triple dividend (Categories 1 and 2)[3]:

1. Benefits to the business undertaking the investments (second dividend captured by the firm itself)
    - Improved business image (from being a "good citizen")
    - Improved credit rating (from increased stability)
    - Improved ability to deal with multiple hazards (from business continuity planning).
2. Benefits to other businesses in the supply chain or geographic vicinity (second dividend captured by connected firms)
    - Increased supply chain stability (from business continuity)
    - Reduction in contagion effects (from lower likelihood of fire spreading or falling debris).
3. Benefits to the general business climate (second dividend captured by other firms more generally)
    - Reduced uncertainty (through lowering the likelihood of disaster losses)
    - Increased economic stability (from business continuity)
    - Increased economic growth (from business continuity)
    - Contributions to technological progress (from embodied technological improvements).

[3]Recall that the first dividend relates to the direct effects of DRM in reducing disaster risk to the business itself, so is not a co-benefit.

4. Benefits to society (second dividend to the broader society or ecosystem)
   - Improved health and education (from employee-related measures)
   - Improved environment (from more prudent use of resources).
5. Joint product benefits (third dividend)
   - Increased productive capacity in addition to resilience capacity
   - Increased production of ordinary goods and services.

5.2.5 Bounded Rationality

Economics has long been criticised for invoking simplifying assumptions about behaviour, such as the profit maximisation motive of businesses. A broader perspective has evolved that incorporates behavioural considerations. The main body of this approach is known as "bounded rationality", and it focuses on various limitations to decision-making (see, e.g., Girgerenzer and Selten 2002). A classic example is myopia, which refers to the use of unduly short time horizons. The manager may be interested only in near-term gains, as opposed to long-term ownership considerations. Kunreuther and Michel-Kerjan (2011) found analogous myopia was the major deterrent in undertaking DRM by households in hazard-prone areas. Even major information campaigns to raise awareness have been relatively unsuccessful (Kunreuther 2006). A related phenomenon that also leads to market failure is that of asymmetric information. One key example is the "principal-agent" problem, best exemplified at the business level when the manager is not the owner. The manager's incentives may be end-of-year bonuses, often based on maximising sales rather than profits, which is more in line with the efficient allocation of resources.

These instances are pertinent to DRM decisions. A manager may see a mitigation measure as reducing profits in the year in which they are made, without considering the longer-term view that this will reduce disaster losses in the future. Myopia and split incentives are likely to lead to even less attention to broader societal benefits of DRM, such as poverty alleviation, economic stability and sustainable growth. Thus, just demonstrating the existence of these broader co-benefits is not necessarily enough to achieve the desired action. Some remedies that would help promote the pursuit of co-benefits, but not necessarily capturing them all, include working with owners rather than managers, appealing to the reputation of the firm and giving priority for disaster assistance to those firms that cooperate in DRM.

Another limitation of decision-making is lack of information and inability to process the information available. This concept extends to expertise as well. Yoshida and Deyle (2005) found access to expertise (primarily from engineers, insurance managers and consultants) to be the major determinant of small business investment in hazard mitigation. The emergence of the business continuity industry should help in this regard, especially for smaller firms that lack in-house expertise. This is even truer for homeowners, but programmes like the Institute for Business and Home Safety FORTIFIED Program in the US, which links expertise with higher standards, is a promising approach.[4]

5.3 Shortfalls in Private Sector Investment from the Standpoint of Society

5.3.1 Externalities

The concept of "externalities" has grown in prominence as communities, regions and the entire globe have become increasingly interdependent. An externality refers to an action by one individual or business that affects another without being transmitted to the marketplace. Externalities can be either positive or negative, with the negative externality of pollution being the most widespread example (Weimer and Vining 2005).

Many externalities lead to or make society more vulnerable to disasters. These include the generation of greenhouse gases (GHGs) that cause climate change, wasting of water that exacerbates drought or tilling practices that contribute to floods. Clearly, reducing negative externalities that make society more vulnerable to disasters should be a target of DRM. For example, if decision-makers are convinced climate change will cause drought and increased frequency and magnitude of hurricanes, they can consider mitigating the root cause (the emission of GHGs) rather than waiting for and mitigating the disasters that climate change will spawn. Another reason to do so is to prevent the irreversible damages that even this subsequent mitigation cannot prevent.

Furthermore, some mitigation measures generate negative externalities, which are essentially "co-costs" or negative co-benefits. For example, in Mexico (ODI 2014) a major flood mitigation project resulted in water being diverted, such that flooding was merely "transferred" to another populated area. The case of DRM can be strengthened if these unintended consequences can be minimised cost-effectively.

Contagion effects are a subset of externalities that focus on cascading interdependencies between entities. Examples relate to the spread of fire among adjacent buildings or of disease across the population. This means a business that installs a sprinkler system or contains the drainage from its own property will benefit its neighbours. Unfortunately, it is unable to reap the entire reward of these actions, as the next subsection discusses more thoroughly.

---

[4]"FORTIFIED Program", accessed 2 February 2016, http://www.smarthomeamerica.org/fortified.php.

Overall, Brugmann (2012) and others emphasise the importance of "internalising" various externalities, including the co-benefits of DRM, into investment calculations. Brugmann notes, "The challenge of climate adaptation, and of other risk reduction strategies, is to create the institutional, planning and policy frameworks, business practices, information systems and financing instruments to establish a market basis for resilience upgrading of vulnerable urban areas and systems as private investments flow into these assets and areas over the coming decades" (2012, 217–18). Of course, the application of more standard approaches to internalising externalities should also be considered, including regulations, taxes and subsidies. These include local zoning ordinances to restrict growth in flood zones, carbon taxes to stem climate change and subsidies to entice behaviour that generates broader social benefits of DRM.

5.3.2 Public Goods

The concept of "public goods" is especially pertinent to DRM. A public good is one (1) that can benefit more than one entity at a time without detracting (much or at all) from the benefits of another and (2) from whose benefits exclusion is technologically infeasible, socially unacceptable or economically unviable. It is not actually the same thing as a project having multiple benefits, but focuses on the tenor of those benefits. Thus, the construction of a hydroelectric dam generates a private good (electricity) and several public goods (e.g. flood control and recreation). Because entities cannot always be excluded from partaking in the benefits of the good, they cannot always be charged a price for it. Hence, those goods with publicness attributes will be underprovided in the marketplace. It is not just the usual expense, or "lumpiness", as it is sometimes referred to (see, e.g., Vorhies 2012), but also this combination of widespread benefits and the inability of individual businesses to capture them that lead to this type of "market failure".

At the same time, public goods need not always be provided by government. Education has public good characteristics—the same schoolroom being able to serve many students and it being socially unacceptable to deny students an education. However, private schools also flourish, even though they are not likely by themselves to lead to the socially optimal level of education (for one thing, they exclude the masses of people who cannot afford schooling). Large-scale projects, such as major hydroelectric dams, are almost always provided by the public sector, but lesser DRM tactics, even those that involve some public goods characteristics, can be provided by the private sector with some government institutional or regulatory support. In addition, subsidies, such as tax rebates, can be provided to promote the level of a public good in the best interests of society, by making up the gap between the public and private optima.

5.3.3 Economic Interdependencies

No business operates in isolation; it requires inputs from its suppliers and sells its products to other businesses, governments or households. In essence, the economy is a set of integrated supply chains. These "economic interdependencies" can cause losses to one entity to ripple throughout the economy. At the national level, this is often measured in terms of a multiplier (the ratio of total impacts to direct impacts), with numerical values two to four times the initial hit. However, multiplier effects can be even larger when we consider critical material shortages or key bottlenecks in the economy that disasters can cause.

Thus, the lack of adequate DRM by one business can negatively affect the entire economy if substitutes for goods or services cannot be found or produced, or only at a higher cost. One would not expect businesses to actually value these implications for the broader community, but they should be taken into account (see also Chaps. 2 and 3). Hence, only government is likely to be able to incorporate this aspect into DRM decisions (see Chap. 4 for discussion of fiscal entry points).

Interdependencies can be broadened to include many categories of indirect economic losses. Hallegatte and Przyluski (2010) stress that direct losses from disasters are not sufficient to garner support for DRM. Other forms of interdependencies include losses stemming from interdependent infrastructure systems, such as those that led to the Fukushima disaster, or cascading failures in general, as evidenced by Hurricane Katrina.

5.3.4 Moral Hazard

The concept of "moral hazard" refers to engaging in negligent actions because the entity need not bear their full cost. A classic example is in the area of natural hazards, when businesses and households continue to rebuild in floodplains because they are certain they will receive continued government assistance after the next flood takes place.

Moral hazard poses a classic dilemma for government. Those hit by a disaster are the neediest in society, and it is difficult to refuse aid immediately following a crisis. In other cases, post-disaster assistance is a means of currying favour with certain key constituencies. Several obvious remedies exist to reduce moral hazard, including stating in advance that post-disaster aid will be reduced; land-use planning that forbids development in the floodplain; and buying out of structures in hazardous areas (the MMC study (2005) noted above found floodplain buyouts to have some of the highest cost–benefit ratios).

Many well-intended policies contribute to the moral hazard situation. For example, the Ministry of Finance and Public Credit in Mexico has implemented two major strategies: (1) creation of a natural disaster fund and (2) risk transfer through reinsurance and catastrophe bonds. The former sends a signal to potential disaster victims that they will be bailed out. The latter does not really reduce disaster risk but simply spreads it out. This is not to say the policies are not well intended or beneficial, it is just that they have some features that do not reduce overall risk. At the same time, there are several ways to stem moral hazard, such as the deductibles in insurance policies or "index-based" insurance, where protection is gauged in terms of factors that cause losses rather than the level of losses.

5.3.5 Distributional Considerations

Often neglected in risk assessments is the distribution of costs and benefits. This is important for at least three reasons. First, numerous studies have determined that the worst-off groups are those most vulnerable to disasters—and disaster exacerbates their condition. Thus, disasters are a great concern from an equity, or fairness, standpoint. Second, lagging social economic groups or regions have been found to represent a drag on economic growth and development. Third, identifying the impacts on various stakeholders provides insights into the motivations of government decision-makers and the likelihood of support or lack of support for DRM policies. Distributional information can better inform stakeholders and thus enhance the public participation process, as well as serving as a predictive tool for the decisions the process is likely to yield. Used appropriately, distributional data can fill in many informational gaps and help lead to a more enlightened citizenry, and hence DRM more attuned to the needs of the public (Rose et al. 1989).

5.4 Co-benefits of Public Sector Investment

5.4.1 Co-benefits to Society

Not adequately considering the co-benefits of DRM is a major reason for under-investment (Vorhies 2012). Public officials in general need to be better informed of these co-benefits (in this case both the second and the third dividends) and to do a better job of communicating them to their finance ministries and to the business community.

To expand on the discussion in the previous section, there are several reasons why even the knowledge of co-benefits may not be sufficient for adequate public sector action. Such reasons come under the heading of "government failure", the counterpart to "market failure". One of them is the short time horizon of elected officials, who often cannot see beyond the next election. A counterpart is the insulation of appointed officials. Further exacerbating the problem is the opportunity cost of DRM investment in relation to other goods and services that public officials may prefer to provide, as they view the public interest or their own political self-interest (Vorhies 2012). There are several ways of overcoming this problem.

For example greater public participation in the decision process injects a two-way flow of information that promotes DRM. Flores and Smith (2010) note that democratic societies tend to do a better job in terms of DRM.

5.4.2 Co-benefits to the Private Sector

Despite an exponential increase in financial losses from disasters, businesses have not kept pace in terms of spending on DRM. The issue is how to engage the private sector in DRM from a finance and implementation standpoint. Also important is how this engagement would influence government development planning.

One more direct and apparent benefit to the private sector relates to risk-taking. Analysts have long noted the important role of risk-taking in entrepreneurship, and it should not be stifled but rather enhanced. Investors work in the context of background risk not under their control, but disasters exacerbate this and can lead to greater risk aversion, which has a dampening effect on the entrepreneurial spirit. Reducing this background risk and providing better information on residual risk can help promote DRM (second dividend). It leads to a context of "risk-conscious decision-making" (see Chap. 2).

There is an important role for insurance in DRM. Throughout most of the industrialised world it is the private sector that provides insurance, and its expansion into the developing world would be valuable. However, initiatives are likely to require backing through reinsurance, government subsidies or government regulation (e.g. ceilings on liability/pay-outs). One of the especially beneficial aspects of insurance, when structured properly, is the inducement for mitigation. This harkens back to fire loss coverage, whereby the insured are given reductions in premiums if they undertake sounder fire prevention practices. Lower insurance rates represent an additional co-benefit of DRM, in that it makes insurance more affordable to a larger number of people.

5.5 The Resilience Dividend as a Sustainable Development Theme

Awareness of the co-benefits of DRM must be raised. Because so many of the co-benefits are not part of individual business enterprise calculations, they are likely to be ignored. Also, because many of them are especially difficult to measure, their full contributions may not be fully appreciated. Knowledge transfer can be especially valuable in making the business case for DRM. Best practice methods of evaluation, instruments for finance and techniques for implementation can be very valuable in this regard. This should also be extended to what is referred to as "next practices", which are more forward-looking and help developing countries overcome some of the mistakes and limitations of the past by learning from industrialised country experiences.

Resilience is considered the response to short-run shocks, whereas sustainability refers to maintaining a long-run development path in which actions by the current generation do not compromise the wellbeing of future generations, with an emphasis on maintaining the value of natural capital and human capital, not just physical capital. If a country cannot survive short-run shocks, it clearly will not be sustainable. One other aspect of the relationship is key: transforming the ingenuity that arises from short-run resilience into longer-term practices to promote sustainability. Rose (2014) specifies several steps that can help in achieving this goal. This pertains primarily to the second dividend, but also to the third dividend.

5.5.1 No Regrets Strategy

One of the major features of the co-benefits approach to evaluating DRM investments is that many are not dependent on actually experiencing the disasters they seek to prevent. Sometimes such investments might be viewed as wasteful, but this is an unfair characterisation, since from a probabilistic standpoint they are prudent. Thus, while the expected value of benefits (losses averted) is positive, the actual direct benefits may be viewed as zero. However, the co-benefits often take place irrespective of the occurrence of any disasters. It has been popular to refer to such instances as a "no regrets strategy", in that it reaps benefits irrespective of future outcomes. This terminology has come into widespread use in the climate change area, where tactics such as energy efficiency (reductions in energy use that more than pay for themselves) are considered meritorious on their own even if predictions about climate change are not accurate (IPCC 2014). Hence, the connection here is to the first dividend.

Clearly, pure no regrets strategies are just a subset of DRM investments where co-benefits themselves outpace the costs. As such, they relieve the pressure on some highly visible projects that do not otherwise appear to have been needed. We should, however, not let the perfect be the enemy of the good, in that investments with partial offsets of costs through co-benefits are worth seeking on a probabilistic basis.

5.5.2 Shared Growth and Social Benefits

Prudent DRM investments will not only increase the profit margins of firms but also benefit the entire economy directly and indirectly. Direct benefits stem from the increased capital stock and production (contributions to a higher GDP) of the firm itself. Indirect effects, or co-benefits, stem from the many categories discussed in this chapter, including multiplier effects, employment opportunities and tax revenues in all cases, and environmental and broader social benefits in some cases.

Moreover, the reduction of uncertainty can have a stimulating effect on both the firm and the overall economy through many conduits, including attracting more foreign direct investment (second and third dividends).

Most of the population can share the gains of economic growth. Increased employment helps reduce poverty and provides improved health care for those employed (again, directly and indirectly). Increased tax revenues can be used to help others, not only in terms of health care but also with respect to education and other social services. Reduced uncertainty provides broader social benefits as well in terms of locational choices, personal investment planning and human resilience.

There is some controversy about which socioeconomic groups benefit most from DRM. The well-to-do benefit because they have the most assets. On the other hand, the poor often live in areas with greater hazard exposure and vulnerability. Moreover, because they live on the edge of subsistence, even relatively smaller losses can be relatively more injurious to them. The economic and social equity implications of DRM need to be examined and refinements made where they violate a society's principles of fairness. This is not an easy matter, in part because of the diverse relevant equity principles are espoused and sometimes conflict, as when for example comparing the benefits principle with ability to pay at the local level and the many alternative burden-sharing principles for mitigating GHGs (see, e.g., Rose 2009b).

5.5.3 Environmental Benefits

Climate change has placed a much stronger focus than ever before on the role of the environment in disasters. Previously, only a very small portion of DRM had been oriented towards the environment (third dividend). For example, a major study of the benefits of 10 years of US Federal Emergency Management Agency (FEMA) hazard mitigation grants (Rose et al. 2007) indicates that, while significant portion of the grants yielded environmental benefits, especially those associated with flood mitigation, less than 1 % of the total dollar benefits were environmental. The small fraction resulted despite the broad range of benefits, which included improved water quality (for recreational and commercial fishing, drinking water), reduction of hazardous waste and enhancement of wetlands and aesthetic and health and safety benefits. In hardly any of the cases were environmental benefits cited as a major concern, but they were factored into the FEMA grant applications as co-benefits. At the same time, it should be mentioned that this study and others underestimate environmental benefits because of measurement difficulties in this realm.

In the case of climate change, not only is the environment is the medium through which the disaster is transmitted but also aspects of it are among the major receptors of damage. Threatened on a much broader scale than ever before are delicate ecosystems, biodiversity and soils, among others. Projected economic losses from climate change are in the hundreds of billions of dollars per year, with the sizable portion being environmental (IPCC 2014). Abatement of GHG emissions is being justified on the basis of all of these losses, and adaptation to climate change manifestations can further reduce losses. Most of the improved environmental areas are public goods or common property resources, but still many of them are utilised by the private sector (e.g. river water for cooling, pristine areas for recreation). Many environmental goods and services have private goods characteristics as well and are of course directly related to private sector interests. These include timber, soils and biodiversity, which surface temperature warming, drought and increased wind and flood damage are threatening more than ever. In these cases, environmental services are direct benefits rather than co-benefits.

5.6 Incentivising Policies

5.6.1 Financial Instruments

Paying for DRM is a critical consideration. Many businesses, especially those in developing countries, do not have the necessary internal resources at their disposal. Several means of private financing for DRM have been developed and have been shown to be effective, although they are not without their limitations. Still, "disaster financing" often refers to risk-spreading and government or philanthropic transfer payments rather than outright risk reduction, although the latter can be incorporated into the former through innovative policy design.

Innovative financing instruments include: land leases, land exchanges, "bonusing" incentives and value capture. These are in addition to standard private sources such as grants, insurance/reinsurance, securitisation, performance contracts, custom debt instruments, equity, guarantees and loans (Brugmann 2012, 224–5).

Credit (business loans) is one alternative, but it is not unusual to find it available only from private sources at higher than market interest rates, given the desperation of borrowers in post-disaster situations. Moreover, unless appropriately devised, this instrument can lead to moral hazard, thereby exacerbating losses. Clearly, there is a role for government provision or regulation of credit in these situations. For example, some Latin American countries, where microcredit falls short, have developed public–private partnerships for what is called an Emergency Liquidity Facility.

Securitisation is an approach whereby a stable and predictable revenue stream exists. It essentially represents a claim on these future earnings and can be sold to obtain capital at the beginning of the earnings stream. This is a popular private sector instrument, and the revenue stream could be diverted for DRM or the instrument used more explicitly to fund DRM investments.

Another major approach is insurance. In the business sphere, this is a valuable contributor because the outright absorption of catastrophic losses would lead to the demise of all but the largest and strongest enterprises. The disaster insurance business itself is in an especially precarious position given the magnitude and geographic concentration of claims, and typically requires a safety net of its own in the form of reinsurance or government backing. Covered losses are still only a fraction of total property and casualty losses from disasters—nearly 50 % on average in a country like the US but a very low percentage in developing countries. One attribute of insurance that can be of immense help is the actuarial information it provides in assessing risk. In this sense, insurance agents can serve as effective risk messengers.

Another type of insurance can also play an important role, although its prevalence today has been confined primarily to industrialised countries. This is ordinary business interruption insurance, which covers lost profits when a business cannot operate owing to damage to its own facility, or contingent business interruption insurance, whereby the business is unable to operate because one of its suppliers (including utility lifelines) or employee access is disrupted (Rose and Huyck 2016). This form of insurance provides working capital needed to purchase inventory for resale or inputs for production, over and above standard property and casualty insurance policies, which cover repair and reconstruction.

A related instrument, sometimes referred to as "quasi-insurance", is the catastrophe (CAT) bond. This is used to spread extreme risks, whereby investors receive a premium in exchange for the bonds but forfeit the principle if a disaster strikes. Bond revenues represent a source of private capital that can be used for mitigation or building resilience capacity instead of after-the-fact recovery and reconstruction. Obviously, the upfront use of funds has a great deal of merit but has been underused. CAT bonds are increasingly being used by local government authorities, but could also be adopted by the private sector (Michel-Kerjan et al. 2011).

Swiss Re (2011) discusses various other refinements of insurance products, as well as successful real world examples. Also, insurance, through rate reduction incentives, represents one of the most successful ways of promoting mitigation. Still, there is a long history of under-adoption of insurance (Kunreuther et al. 1978), although lessons learnt and research innovations are helping improve design (Kunreuther et al. 2013). For example, Kunreuther and Michel-Kerjan (2010) have proposed combining multiyear property insurance with loans to promote mitigation that can reduce premiums.

The investment industry is likely to find various co-benefits of DRM appealing if they can be quantified. This includes bond-rating companies that seek to reflect asset risk in relation to the ability of bond issuers to make payments. It also includes investment companies, and mutual, hedge and pension funds and others that have a long-term fiduciary responsibility to protect investments (MMC 2015).

5.6.2 Regulation and Government Incentives

Regulation is likely necessary in risk financing for several reasons, mainly guarding against exploitation of the vulnerable. This applies to capping loan rates, setting insurance rates and promoting good faith adjudication of claims. Government can also promote the adoption of these instruments through improved risk communication, including consideration of co-benefits. Swiss Re (2011), for example, views government (and non-governmental organisations) as valuable partners to the insurance industry. Among roles it can perform are rule setter, sponsor for insurance markets, reinsurance supplier/backer and reinsurance purchaser. Also, an insurance mandate is one of the most powerful ways to generate the co-benefits of avoiding adverse selection and to cross-subsidise insurance.

As noted above, various types of subsidies might be warranted to induce the private sector to take on DRM that benefits the broader community, with the subsidy being set at the level needed to bring about the socially optimal (additional) level of DRM. Tax breaks have been very effective at doing so in the area of energy conservation in many countries, and this approach holds promise for DRM. In fact, there is the possibility to tie some energy-related aspects of DRM to these existing tax reduction incentives.

Penalties can also be used to discourage behaviour that increases risk, such as locating in hazard-prone areas, placing undue burdens on infrastructure and releasing toxic materials. One approach to funding DRM stems from charging for negative externalities. This has become more prevalent with greater identification and measurement of damages. Examples include utility price "adders", pollution taxes and the auctioning of tradable emission permits. In the latter two cases, the government can use revenues to promote worthy goals, including the mitigation of negative externalities in the first place and promoting economic equity, as in providing tax relief and in-kind or cash transfers for those most adversely affected, especially lower-income groups (see, e.g., Rose 2009b).

5.6.3 Broader Approaches

Long-run considerations have for decades been increasingly incorporated into decision-making relating to business environmental management, and more recently have been integrated into sustainable planning. In the US, more than 300 of Standard and Poor's 500 Stock Exchange Index businesses report their GHG emissions. The Brazilian firm Natura chooses suppliers on the basis of economic, social and environmental performance. Alcoa bases 20 % of its executive compensation on progress on sustainability (Perera et al. 2013). Moreover, non-governmental organisations, such as the World Resources Institute, have worked with the private sector to evaluate and improve environmental performance. These initiatives in the environmental area could be extended to DRM, especially in light of recent findings in general and risk management. A survey by the Economist Intelligence Unit (EIU) (2007) found that protecting and enhancing the firm's reputation was ranked even higher than preventing financial loss. Still, terrorism and natural weather events ranked eighth and ninth as the most important external drivers of improvements and risk management in the EIU survey of firms. The study found that the major hindrance to action in DRM was the difficulty in measuring its benefits, a finding that makes improving measurement all the more important in promoting DRM.

Another broader perspective is to make disaster resilience, including co-benefits, an integral part of asset management in business and community operations (MMC 2015). Of course, it takes time to institutionalise such advances but increasing the awareness of DRM co-benefits can help.

A new "layered" analytical approach to risk-framing distinguishes major categories to improve the targeting of DRM. Linnerooth-Bayer and Hochrainer-Stigler (2015) describe three layers:

1. Frequent but low-consequence events. Here, risk reduction is often cost-effective.
2. Rare but high-consequence events. Here, risk finance is usually most appropriate.
3. "Beyond adaptation." This refers to the realisation that some risks, especially those associated with the spectre of climate change, are so great as to defy risk management from either a technical or an economic liability standpoint. Hence, government or donor aid is warranted. There is also the implication that some future disasters may be even beyond these capacities.

Brugmann (2011, 2012) promotes a holistic approach that integrates climate change adaptation and other DRM-related objectives into broader sustainability planning, taking account of the many interdependencies in urban areas. The spectre of severe damage from climate change and the shift in focus to adaptation has resulted in a reframing of the financing for DRM. Brugmann (2012) characterises this as a shift away from disaster risk reduction as the end goal to a sustainable development focus on financing the performance of urban assets, areas and systems.

It is well known that the riskiest urban places manage to attract investment. However, most analysts believe public investment is not up to the task of mitigating the increasing risks on the horizon. Brugmann proposes leveraging funds for capacity-building and internalising the downsides of future development so that vulnerabilities become *investment opportunities*.

Brugmann (2012) also notes that his "resilience upgrading" concept alters investment strategy from an international project finance approach to a more diversified and leveraged approach in a market context by incorporating a "value proposition" at the outset, which is better able to capture private investment. He refers to this as "mainstreaming" DRM and sustainability investments into development planning, which he claims have relatively lower risk than special purpose investments (Green buildings are offered as a major example of an investment category that has been mainstreamed through performance contracts that have resulted in demonstrations of more stable and higher returns on investment).

5.6.4 Imperatives

Climate change poses serious challenges because of the difficulty of evaluating the risk of new and distant threats. Surminski (2010) has summarised efforts by insurers, mainly in industrialised countries, to finance mitigation strategies in this context. Linnerooth-Bayer and Hochrainer-Stigler (2015) note that most of these attempts have been ad hoc and in the form of improved risk communication at the micro level. They encourage new approaches, including loss modelling to improve economic assessments.

Additions to the current level of DRM, especially in developing countries, are imperative. At present, losses are often absorbed by governments or donor institutions, whose coffers will be even more strained as disaster losses increase (under business-as-usual projections). Stronger efforts are needed to promote DRM in the form of mitigation, preparedness, post-disaster resilience and long-term adaptation. Other approaches are needed as well, such as the increased transfer of risk to capital markets. Risk finance and risk reduction are not necessarily mutually inclusive, and it is important to capitalise on the complementarities between these two strategies.

Unfortunately, DRM will have to be undertaken in the context of a new reality in many countries of Europe, the US and Japan, where aging populations are slowing the growth of the economy and government tax revenues (Pisano and Callahan 2012). This has inspired a trend among local public officials to link fiscal and environmental decisions, which has essentially made officials in several instances more crisis- and sustainability-conscious, thereby promoting the consideration of longer planning horizons (Callahan and Pisano 2014). It has also resulted in formulating environmental sustainability planning in a manner that demonstrates its worthiness in terms of performance, including considerations of return on investment. Callahan and Pisano (2014) emphasise the need to develop a constituency for fiscal sustainability, as has been done for its environmental counterpart. Means to this end are improved communication with beneficiaries, including identification of what we refer to as co-benefits. It would also extend to training to increase capacity to value DRM benefits and to identify and evaluate actions to promote it.

The presence of co-benefits helps justify various types of expenditures on DRM at several levels. Those that are internal to the firm can be counted as part of a broader set of returns on investment. Those that are external to the firm can be addressed in several ways. One example is collective action by parties that benefit. Governments can play a role here in coordinating efforts and subsidising some part of the DRM for the common good. Even areas of risk finance, specifically with respect to governments or private donors providing assistance, can be rationalised by the existence of co-benefits. What is often considered bailing out individual businesses is justified in part because of the broader contributions to such objectives as economic stability, growth and sustainability.

5.7 Conclusion

Co-benefits refer to DRM investment outcomes other than those intended to directly lower risk from disaster. They occur under the first and second dividends of our framework.

Most businesses maximise short-term profits and will only invest in amounts and the manner of DRM that promote this goal. More enlightened firms have broader goals, such as maximising long-run profits, and this may make them more likely to spend more on DRM. However, businesses often do not even invest sufficiently in DRM to maximise either short- or long-run profits, given endemic problems such as the "principal-agent" situation of large firms, in which the motives of managers differ from those of owners.

There are several recipients of co-benefits:

- The business undertaking the investments
- Other businesses interacting with the firms undertaking the investment
- The general business climate
- Society.

Co-benefits of DRM offer several attractive rewards to businesses. Some co-benefits are more certain, tangible and immediate than the intended direct benefits of mitigation, which may never materialise if the disaster does not take place. Some are consistent with a no regrets strategy in that they yield cost savings irrespective of whether a disaster strikes. Some DRM investments provide protection against several hazards in addition to those that are the main focus. Others contribute to shared growth of the economy and to broader social benefits. Some represent payment for unpriced services, such as the environment; these payments enhance the reputation of businesses and can thus serve to increase long-run profits.

This chapter has presented the case for private sector consideration of the co-benefits of DRM. We have elucidated the co-benefits of private DRM and explained the conditions under which they are and are not considered. We have also examined the relationship between private and public sector aspects of the issue. We have explained how broadening the picture of resilience as an important contributor to sustainable development would help identify additional co-benefits and provide a decision-making framework in which these would be more fully appreciated. We have also established the basis for formulating incentives and other means of pursuing DRM goals within the private sector beyond its pursuit of more narrow goals.

Co-benefits of DRM would seem to be subservient to direct benefits, especially in the private sector decision-making process. However, neither the direct benefits nor the co-benefits of DRM typically result in revenue flows, and thus they are both relegated to lesser importance than investments having an impact in terms of increases in sales revenue (or decreases in production costs). Thus, it would help to estimate co-benefits in monetary terms, so they can be viewed in the light most businesses understand.

The post-disaster phase, for which we have reserved the term "resilience", is especially critical for businesses in relation to their survival. This is a strong inducement for them to operate efficiently and recover quickly (recall our static and dynamic economic definitions of resilience). However, this does pose some problems for broader perspectives on DRM. Businesses are likely to be especially inward-looking and short-sighted during recovery, out of what they perceive as necessity. Longer-term considerations and broader societal benefits may be pushed aside. Co-benefits are less likely to fit into the decision-making process at this stage. Moreover, even longer-term opportunities to reduce vulnerability through enhanced mitigation and the building of resilience capacity ahead of the next disaster will likely be considered of secondary importance. This may signal an even stronger need for government involvement in DRM at this stage to identify and help induce investments in the best long-term interests of society as a whole.

Incentivisation of DRM for the private sector faces some serious obstacles. Disaster risk is a relatively minor consideration in many areas, such as construction loans, although it is a larger consideration for property appraisal. Insurance policies are property-specific and do not consider the condition of the community in its entirety, which reaps the greater share of co-benefits. Also, insurance premiums are such a small percentage of business costs that it is easy to overlook an even smaller incentive rate reduction. At the same time, globalisation has led to even greater competitive pressures to consider all benefits, as well as costs, and to greater capital mobility, which has increased pressure to reduce investment risk.

Unfortunately, examples of private sector consideration of a comprehensive set of co-benefits of DRM are limited. It is hoped this chapter will help in identifying opportunities, as well as obstacles and ways to overcome them, to take co-benefits into account in private sector investment decisions on DRM. In addition, it is intended to provide insight into ways policy-makers can facilitate private sector actions by removing obstacles and providing incentives.

References

Alberto AJ, Sharma S (2003) A contingent resource-based view of proactive corporate environmental strategy. Acad Manage Rev 28(3):71–88

Brugmann J (2011) Financing the resilient city: a demand-driven approach to development, disaster reduction and climate adaptation. White paper, ICLEI, Bonn

Brugmann J (2012) Financing the resilient city. Environ Urbanization 24(1):215–232

Callahan RF, Pisano M (2014) Aligning fiscal and environmental sustainability. In: Mazmanian DA, Blanco H (eds) Elgar companion to sustainable cities: strategies, methods and outlook. Edward Elgar, Cheltenham

Carrera A, Mesquita L, Perkins G, Vassolo R (2003) Business groups and their corporate strategies on the argentine roller coaster competitive and anti-competitive shocks. Acad Manage Executive 17(3):32–44

Doh JP, Pearce JA (2004) Corporate entrepreneurship and real options in transitional policy environments: theory development. J Manage Stud 41(4):645–664

EIU (Economist Intelligence Unit) (2007) Best practice in risk management: a function comes of age. The Economist, London
Field C, Barros V, Stocker T, Dahe Q, Dokken DJ, Ebi KL, Mastrandrea MD, Mach KJ, Plattner GK, Allen SK, Tignor M, Midgley PM (eds) (2012) Summary for policy makers. In: Special report on managing the risks of extreme events and disasters to advance climate change adaptation: summary for policymakers. Report of working groups I and II of the IPCC. Cambridge University Press, Cambridge and New York
Flores AQ, Smith A (2010) Surviving disasters. Wilf Family Department of Politics, New York University, New York
Girgerenzer G, Selten R (eds) (2002) Bounded rationality: the adaptive toolbox. MIT Press, Cambridge
Hallegatte S, Przyluski V (2010) The economics of natural disasters: concepts and methods. Policy research working paper 5507, World Bank, Washington, DC
Hoffmann VH, Trautmann T, Hamprecht J (2009) Regulatory uncertainty: a reason to postpone investments? not necessarily. J Manage Stud 46(7):645–664
IPCC (Intergovernmental Panel on Climate Change) (2014) Fifth assessment report (AR5). IPCC, Geneva
Kull D, Mechler R, Hochrainer-Stigler S (2013) Probabilistic cost-benefit analysis of disaster risk management in a development context. Disasters 37(3):374–400
Kunreuther H (2006) Disaster mitigation and insurance: learning from Katrina. Ann Am Acad Polit Soc Sci 604:208–227
Kunreuther H, Ginsberg R, Miller L, Sagi P, Slovic P, Borkan B, Katz N (1978) Disaster insurance protection: public policy lessons. Wiley, New York
Kunreuther H, Michel-Kerjan E (2010) From market to government failure in insuring U.S. natural catastrophes: how can long-term contracts help. In: Brown JR (ed) Private markets and public insurance programs, American Enterprise Institute Press, Washington, DC
Kunreuther H, Michel-Kerjan E (2011) Redesigning flood insurance. Science 333(6041):408–409
Kunreuther H, Michel-Kerjan E, Pauly M (2013) Making America more resilient toward natural disasters: a call for action. Environment 55(4):15–23
Linnerooth-Bayer J, Hochrainer-Stigler S (2015) Financial instruments for disaster risk management and climate change adaptation. Clim Change 133:85–100
Michel-Kerjan E, Zelenko I, Cardenas V, Turgel D (2011) Catastrophe financing for governments: learning from the 2009–2012 multicat program in Mexico. Working paper on finance, insurance and private pensions 9, OECD, Paris
Mitchell T, Van Aalst M, Villanueva PS (2010) Assessing progress on integrating disaster risk reduction and climate change adaptation in development processes. Institute of Development Studies, Brighton
Multihazard Mitigation Council (MMC) (2005) Natural hazard mitigation saves: an independent study to assess the future savings from mitigation activities. Report to U.S. Congress on behalf of the National Institute of Building Sciences, Washington, DC
Multihazard Mitigation Council (MMC) (2015) Developing pre-disaster resilience based on public and private incentivization. Washington, DC
(ODI) Overseas Development Institute (2014) Case studies to support realising the resilience dividend, climate and environment Programme. ODI, London, UK.
Perera A, de Pino SP, Oliveira B (2013) Aligning profit and environmental sustainability: stories from industry. Working paper, World Resources Institute, Washington, DC
Pisano M, Callahan RF (2012) Case study I: fiscal sustainability in Los Angeles County. Nat Civic Rev 101(1):11–17
Rose A (2009a) Economic resilience to disasters. Community and Regional Resilience Institute Research Report 8. Oak Ridge National Laboratory, Oak Ridge, TN
Rose A (2009b) The economics of climate change mitigation policy: international, national and regional perspectives. Edward Elgar, Cheltenham

Rose A (2014) Economic resilience and the sustainability of cities in the face of climate change: an ecological economics framework. In: Mazmanian Daniel A, Blanco Hilda (eds) Handbook of sustainable cities. Edward Elgar, Cheltenham, pp 336–353

Rose A, Huyck CK (2016) Improving catastrophe modeling for business interruption insurance needs. Risk Anal, forthcoming

Rose A, Stevens B, Davis G (1989) Who gains and who loses from natural resource policy: distributional information and the public participation process. Resour Policy 15(4):282–291

Rose A, Oladosu G, Lee B, Asay GB (2009) The economic impacts of the 2001 terrorist attacks on the world trade center: a computable general equilibrium analysis. Peace Econom Peace Sci Public Policy 15(2) (Article 4)

Rose A, Porter K, Tierney K (2007) Benefit-cost analysis of FEMA hazard mitigation grant. Nat Hazards Rev 8(4):97–111

Sheffi Y (2005) The resilient enterprise. MIT Press, Cambridge, MA

Surminski S (2010) Adapting to the extreme weather impacts of climate change—how can the insurance industry help? Climate Wise

Swiss Re (2011) Closing the financial gap: new partnerships between the public and private sectors to finance disaster risks. Swiss Re, Zürich

Tanner T, Surminski S, Wilkinson E, Reid R, Rentschler J, Rajput S (2015) The triple dividend of resilience: realising development goals through the multiple benefits of disaster risk management. Global Facility for Disaster Reduction and Recovery (GFDRR) at the World Bank and Overseas Development Institute (ODI), London

Vorhies F (2012) The economics of investing in disaster risk reduction. Working paper, UNISDR, Geneva

Weimer DL, Vining A (2005) Policy analysis: concepts and practice, 4th edn. Pearson, Upper Saddle River, NJ

Yang B, Robert N, Backhouse R (2004) Management of uncertainty through postponements. Int J Prod Res 42:1049–1064

Yoshida K, Deyle R (2005) Determinants of small business hazard mitigation. Nat Hazards Rev 6 (1):1–12

Zolli A, Healy AM (2012) Resilience: why things bounce back. Free Press, New York

Chapter 6
Investing in Disaster Risk Management in an Uncertain Climate

Thomas K.J. McDermott

Abstract Climate change will exacerbate the challenges associated with weather variability and extremes in developing countries. As such, it reinforces the development case for investment in disaster risk management (DRM). Uncertainty about how climate change will affect particular locations makes optimal investment planning more difficult. In particular, our inability to derive meaningful probabilities from climate models limits the usefulness of standard project evaluation techniques such as cost–benefit analysis, that attempt to optimise risk–return trade-offs. This chapter offers a simple decision framework that enables policy-makers to identify the particular circumstances under which uncertainty about future climate change becomes critical for DRM investment decisions. Accounting for climate uncertainty is likely to shift the optimal balance of DRM strategies towards more flexible, low-regret-type interventions, especially those that promote 'development-first' or 'risk-coping' objectives. Such investments are likely to confer additional development dividends, regardless of the climate future in a given location. The analysis also demonstrates that climate uncertainty does not necessarily motivate a "wait and see" approach. Instead, where opportunities exist to avail of adaptation co-benefits—for example where DRM initiatives could help avoid locking in future exposure to climate risk—climate uncertainty provides additional motivation for early investment in DRM.

Keywords Climate change · Decision making under uncertainty · Adaptation and development co-benefits · Project appraisal · Investment screening

---

T.K.J. McDermott (✉)
School of Economics, University College Cork, Cork, Ireland
e-mail: thomas.mcdermott@ucc.ie

T.K.J. McDermott
Grantham Research Institute on Climate Change and the Environment, LSE, London, UK

S. Surminski and T. Tanner (eds.), *Realising the 'Triple Dividend of Resilience'*, Climate Risk Management, Policy and Governance,
DOI 10.1007/978-3-319-40694-7_6

6.1 Introduction

How does climate change alter the business case for disaster risk management (DRM) investments? How should policy-makers treat the uncertain information provided by climate models and make efficient long-term climate-sensitive investment decisions when faced with the wide range of possible climate futures that models predict?

This chapter focuses on the role of climate change—and uncertainty in relation to future climate projections—in the decision-making process for DRM investments. It takes the perspective of the individual policy-maker (government or agency) at local or national level, attempting to allocate scarce resources among competing alternatives, while balancing DRM objectives with other development goals, including the long-term sustainability of wealth creation and the need to adapt to evolving climate risks.

At the aggregate level, the implications of climate change for DRM are unambiguous: climate risk (both extremes and variability) will increase for many developing countries under most climate change scenarios. Climate change therefore reinforces the case for DRM investments (e.g. IPCC 2012). At the level of deciding whether or not to adopt a particular DRM project, on the other hand, climate change complicates things. The nature of climate uncertainty is such that the best available forecasts for a particular decision-relevant variable—for example rainfall intensity or sea-level rise—at an appropriate spatial and temporal scale are in many cases characterised by a wide range of possible climate futures, each of which is "non-discountable". In other words, from a policy- or decision-maker's perspective, the possibility that the outcome will be anywhere within that range cannot be disregarded.

This degree of uncertainty presents considerable challenges for the standard project evaluation decision-making tools, such as cost–benefit analysis. The analysis presented here is complementary to recent research that identifies countries that are already facing disaster- (or climate-) related "stress" (e.g. Mechler et al. 2014). Given the array of policy options available to alleviate that stress—from efforts to reduce exposure via better land-use planning or zoning laws (and their enforcement) to improvements in risk-coping capacity—the challenge for policy-makers is achieving the appropriate balance between the at times competing objectives of DRM, development and adaptation. This task is already difficult without the added complexity of an uncertain climate future. However, this uncertainty need not lead to policy paralysis.

This chapter includes a sketch of a decision-making framework that attempts to simplify the process of accounting for the deep uncertainty associated with climate projections and the specific characteristics of different DRM policy options. In particular, this framework enables policy-makers to identify the particular circumstances under which uncertainty about future climate change becomes critical for their DRM investment decisions.

Many DRM investments made today will have long-term implications, both because some DRM projects are themselves long-lived but also because they will influence spatial and economic patterns of development that involve a degree of path dependency (lock-in). DRM investments that take account of climate risk could therefore have potentially important adaptation co-benefits, while avoiding maladaptation risks.

Disasters worldwide have caused damages of almost $200 billion annually over the past decade, up from $50 billion in the 1980s (GFDRR 2015). In addition to these direct economic losses, disasters can potentially have longer-term effects on welfare and on economic growth, for example via effects on investment and the provision of basic economic and social infrastructure and, perhaps most importantly from a development perspective, via their direct human impacts and indirect effects on the formation of human capital.

These impacts and their potentially long-term effects make DRM a first-order consideration for development policy. In spite of this, developing country ministries of finance and planning appear reluctant to invest in such initiatives (see Chap. 1). International aid efforts also tend to prioritise disaster relief and recovery efforts over risk management (e.g. Kellett and Caravani 2014).

One problem is the relatively narrow framing of standard methods for assessing DRM projects in terms of avoided losses. While avoiding direct human and economic costs is clearly the main objective of any DRM strategy, in many cases there may also be wider "development dividends" associated with initiatives primarily aimed at managing disaster risk (Tanner et al. 2015; see Chaps. 1 and 2). Flood protection schemes might for example encourage inward investment by reducing the risk premium associated with a particular location or enabling safe development of locations that are inherently high productivity but vulnerable to disaster. DRM investments, if designed with evolving climate risk in mind, could also help avoid costly maladaptation, which might otherwise threaten the sustainability of development. On the other hand, some DRM initiatives will entail development trade-offs. For example, zoning restrictions on flood plains might constrain the development of desirable locations.

There is also a potential "third dividend" from resilience in the form of co-benefits that accrue from DRM initiatives, regardless of the realisation of actual disaster events experienced in any given period. Risk-coping initiatives—for example social safety nets and better access to financial services—aimed at reducing the welfare impacts of disasters might simultaneously promote productive risk-taking in the form of increased entrepreneurship, innovation and diversification of economic activity. Similarly, improvements in the dissemination of risk information and community-based disaster preparedness schemes, for example, can bring additional benefits in the form of increased community cohesion and better state–society relations (see Chap. 3).

6.2 Adding Co-benefits into the "Mix" of DRM Policy Options

The full spectrum of DRM policies extends beyond the obvious hard infrastructure investments, such as flood barriers, traditionally associated with disaster mitigation. The various DRM policy options might usefully be divided into two distinct categories. On the one hand are attempts to reduce the amplitude of the stimulus or shock (i.e. hazard management). There are numerous alternative (in some cases complementary) elements to this approach, including attempts to reduce exposure, via changes to planning and zoning laws, building codes, etc., and via defensive infrastructure. These are the interventions traditionally associated with disaster risk reduction efforts. The second set of options relates to efforts to improve risk-coping capacity (i.e. risk management)—accepting that some shocks will occur and attempting to minimise the longer-term welfare impacts of those shocks. The latter channel includes better hazard information, early warning systems, emergency procedures and response systems, as well as economic shock absorbers including insurance, credit and social safety nets. This categorisation of policy options is not intended to suggest an "either/or" binary decision for policy-makers. Optimal DRM strategies will no doubt involve a "mix" of policies aimed at both reducing exposure and improving risk-coping capacity.

How does climate change and its uncertainty affect the optimal balance between reducing exposure and increasing resilience? The appropriate balance might depend on both the type and the severity of the risk—what is known as the "risk-layering" approach (e.g. Hallegatte et al. 2010; Mechler et al. 2014). For example, frequent low-impact events might be mitigated through improvements in basic infrastructure (e.g. drainage systems to prevent urban flooding); the impacts of rarer events might be minimised through attempts to reduce exposure, for example by preventing settlement in hazard zones via better public information and zoning or land-use plans; while for the most exceptional, large-scale, events, improved infrastructure and zoning may not be sufficient or economical (e.g. since these might constrain development of productive urban locations). Instead, early warning systems and evacuation plans, combined with support for reconstruction and reinvestment (bearing in mind moral hazard risks), can help avoid the worst human and longer-term economic costs of such events.

This risk-layering approach identifies appropriate DRM strategies for dealing with different risk profiles. But of course the degree of risk (return period) and exposure to that risk are evolving, both in ways that are outside of local policy-makers' control (climate change) and in ways that are amenable to policy action (exposure and resilience). Climate change has important implications not just for the appropriate adaptive responses to various risk layers but also for policies that will partly determine the precise risk layers faced at a given location, and by what populations.

The various policy options mentioned above will differ in their potential to convey development and adaptation dividends as well as other co-benefits, in addition to their primary objective of managing disaster risk. It is also important for policy-makers to consider how existing development trends, and the need to consider long-term adaptation to a changing climate, interact with and shape DRM investments and their likely outcomes. The following section identifies cases where the objectives of DRM, development and adaptation coincide. However, there are also likely to be trade-offs between these at times competing objectives. Such trade-offs also need to be made explicit in any DRM investment decision. Figure 6.1 presents a simple matrix illustrating development and adaptation implications of various DRM initiatives.

The additional development and adaptation dividends identified here provide further motivation for DRM investments, over and above their potential to reduce losses from disasters. Policy-makers apparently tend to perceive DRM investments as representing sunk costs, with little benefit in the case that disaster does not strike (see Chap. 1). The framework developed here emphasises that DRM investments could also confer "sunk benefits"—for example where DRM initiatives could help avoid locking in future exposure to climate risk, preventing potentially costly maladaptation. In such cases, the uncertainty associated with climate change, far from justifying a "wait and see" approach, instead provides further motivation for early intervention to manage risk and build risk-coping capacity (see further discussion in Sect. 6.4).

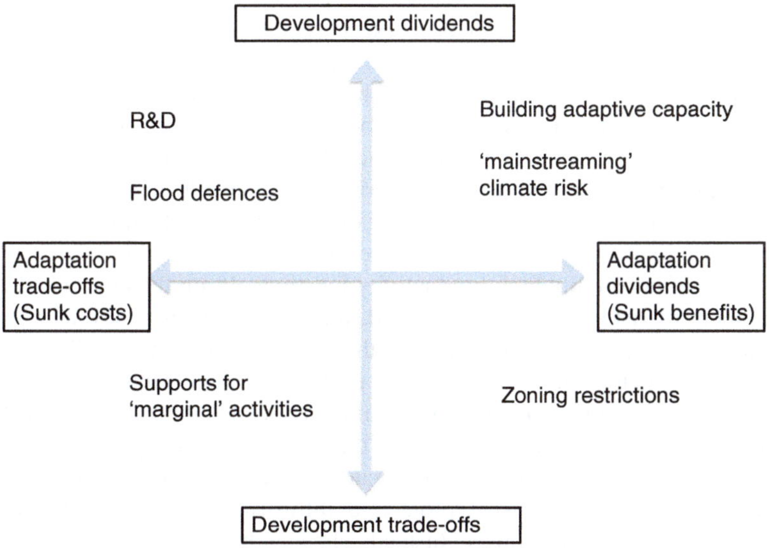

Fig. 6.1 Potential overlaps between development and adaptation dividends (and trade-offs) of various DRM strategies

6.2.1 Development Dividends and Trade-Offs

The debate over DRM investment tends to focus on avoiding losses. But from a development perspective, some losses may cause more harm than others. Large-scale monetary (or asset) losses do not necessarily translate into important consequences for welfare or economic development (Hallegatte 2014). While monetary losses can be recovered through insurance and emergency reconstruction funds, the human impacts of disasters represent permanent losses to welfare.

Economic development necessarily involves risk-taking at the individual household or firm level, in the form of entrepreneurial activities, including experimenting with new technologies, innovation, diversification away from traditional modes of production, etc. The inability of the poor to cope efficiently with risk—and therefore to take on these productive risks—represents one of the essential problems of development (e.g. Bryan et al. 2014). Financially constrained households struggle to cope with risk, employing inefficient coping mechanisms both *ex-ante* and *ex-post* (e.g. Mobarak and Rosenzweig 2013).

Alongside the direct impacts of disasters on human capital, through effects on health (particularly of children) and disruptions to schooling, the threat of disasters also creates a background risk, which may act as an important barrier, for poorer households in particular, to making long-term investments in education. Reducing that background risk generates an important development dividend from DRM investments (see Chap. 1).

But risk reduction should not be pursued for its own sake (Hallegatte 2013). Indeed, one challenge for DRM is that some development trends involve the accumulation of risk. It is for this reason that risk *management* is now emphasised over risk *reduction*. Inevitably, there will be trade-offs between some development and DRM objectives. Obvious examples include trends such as urbanisation and the accumulation of people and assets in at-risk or vulnerable locations, such as on coasts. Restricting such development—for example through land-use planning and zoning restrictions—might reduce exposure to hazards, but might also constrain the exploitation of potentially high productivity locations. If, on the other hand, further development of risky locations is allowed, the question then arises; at what point does increasing exposure to disaster risk threaten development itself?

Hard infrastructure investments, such as flood barriers, might confer a development dividend by reducing the risk premium associated with a particular location or enabling safe development of locations that are inherently high productivity, but vulnerable to disaster, resulting in greater inward investment. However, for developing countries, expensive protective infrastructure may not be the most efficient use of scarce resources. Instead, cheaper but possibly more politically challenging improvements in safety could be achieved through better building regulations and settlement policies (and their enforcement), improved dissemination of information, early warning systems, evacuation and emergency planning and training, etc. Such initiatives would have the benefit of prioritising the protection of human life over economic assets (thus helping minimise welfare impacts), while investing in institutional capacity and community preparedness initiatives might also be expected to have positive co-benefits via improved institutions and governance capacity (see Chap. 1).

DRM strategies that focus on improvements in risk-coping capacity will also promote the adaptive capacity and autonomy of the poor (e.g. by increasing access to financial services and basic infrastructure, etc.). Adaptation should not just be about responding to changes as they occur, but rather efficient adaptation strategies will involve *preparing* to adapt (Oreskes et al. 2010), and in particular building adaptive capacity so individuals, households, communities and nations will have the resources and (economic) flexibility necessary to minimise welfare losses (and maximise welfare gains) from future climate change. Part of such strategies will likely overlap with DRM objectives, potentially leading to improved resilience to disaster and weather risk, greater capacity among the poor for productive risk-taking and capacity-building for future adaptation. Ultimately, such strategies reduce the likelihood of a vicious cycle between risk and poverty.

6.2.2 Adaptation Dividends and Trade-Offs

Many DRM initiatives will overlap with efforts to cope with climate risk. Adaptation to future climate change therefore represents a natural "dividend" of most DRM investments. In particular, initiatives undertaken now to promote sustainable development paths, or to avoid locking in further vulnerabilities, offer potentially large adaptation dividends.

Existing development trends, such as migration to coastal urban areas and settlement in vulnerable locations more generally, increase exposure to extreme weather events. To the extent that these trends are irreversible—for example because of the strong degree of path dependency in urban locations (e.g. Davis and Weinstein 2002)—delaying action to manage those risks could lock in greater future costs to society. Adaptation dividends, particularly in the form of attempts to avoid locking in unsustainable patterns of development, provide a strong motivation for early action on DRM. Delaying adaptive DRM investments likely incurs the opportunity cost of missed opportunities for adaptation. Once a new settlement is established in a vulnerable location, for example, it is difficult to reverse that pattern of spatial development.

In general, adapting to climate change involves trading off opportunities to exploit today's climatic conditions against the ability to exploit (anticipated) future conditions (Millner 2012)—in other words sacrificing some resources today in the expectation of improving future welfare. There are therefore likely to be some trade-offs between adaptation and development co-benefits of DRM. In a developing country context, one might expect the balance of priorities (justifiably) to favour today's challenges and opportunities over (uncertain) future ones. However, this does not imply adaptation co-benefits (and trade-offs) are not relevant factors in the design of appropriate DRM strategies.

Increasing the resilience of rural livelihoods—and in the process avoiding weather risk translating into disasters—involves addressing existing vulnerabilities and improving the capacity of the poor to cope with existing, reasonably well-known and understood risks. As discussed earlier, such initiatives might generate additional development and adaptation dividends above and beyond their primary purpose of reducing the direct losses associated with disasters. The main concern here is to avoid moral hazard or maladaptation in the form of fostering unsustainable modes of production or constraining the opportunities for (and inherent dynamics of) transformative development.

The literature on the resilience of rural livelihoods has tended to focus on in situ forms of adaptation, such as increasing local food security and self-sufficiency (Dercon 2012). While such risk management initiatives would appear to align with adaptation and development objectives, there are risks. For example, Dercon points out that "many drought-resistant crops have low returns, leading to more security but also less poverty reduction". Similarly, in situ adaptation may represent maladaptation where existing activities and locations are already marginal and likely to face deteriorating climatic conditions. Investments in agriculture, such as new seed varieties or irrigation infrastructure, might improve resilience to weather risk in the short term, but could risk locking in forms of production that eventually become unsustainable under climate change (e.g. over-use of groundwater).

Sensible DRM strategies—taking account of the need to consider long-run adaptation to a changing climate—should also consider existing development trends and whether these are likely to be sustainable under a range of possible climate futures, but also how climate change may affect (amplify or diminish) these existing trends. A good example is rural–urban migration. While much of this migration is driven by the "pull" of economic opportunities in urban areas, climate change might also reinforce the "push" of limited economic opportunities and precarious livelihoods in rural areas (e.g. Henderson et al. 2014). On the other hand, climate change also poses significant threats to urban areas in the form of increased risk of heat-waves and flooding. In the absence of adaptation planning and DRM strategies, such threats could lead to the emergence of urban push factors (i.e. a flight from vulnerable urban locations) and the subsequent loss of important development opportunities inherent in high-productivity urban locations.

Climate uncertainty reinforces the need for flexibility—as discussed in greater detail in the subsequent sections. DRM policies should therefore aim to increase the autonomy of poor and vulnerable groups—for example by enabling people to move *on their own terms* (which means moving in some cases, and remaining in place in others). This is likely best achieved through standard development initiatives such as improvements in access to finance and markets (including labour markets) and investments in health, education and basic economic infrastructure (transport, energy and sanitation). In a rural context, then, DRM and development objectives would appear to be reasonably well aligned. DRM considerations, especially under uncertain climate change, reinforce the case for policies that seek to improve the economic flexibility of the rural poor (generating both development dividends and additional co-benefits), but also emphasise the need to consider the long-term sustainability of rural development initiatives.

In an urban context, on the other hand, there is a greater tension between the objectives of DRM and development. Existing development trends (and also potentially rural DRM strategies) will tend to exacerbate urban disaster risk and hazard exposure. Appropriate DRM strategies therefore need to consider both how to support economic transformations (which are an integral part of long-term development) and how to manage the additional risks of large-scale migration to urban areas, which are themselves often vulnerable to disasters, especially urban flooding (Hallegatte et al. 2010).

Defensive investments, such as the construction of flood barriers, may reduce incentives to adapt—a form of moral hazard—or even lead to maladaptation behaviours (e.g. Collier et al. 2008). The presence of a flood barrier will presumably lead either *de jure* or *de facto* to greater development of the protected (but risky) area. In the case of failure, losses would then be exacerbated relative to some baseline scenario. Many hard infrastructure DRM projects (e.g. flood barriers) also face the challenge of providing either complete protection or complete failure. This places an even greater burden on designers and policy-makers to get the level of protection right, and raises again the challenge of dealing with uncertain future risks. Characterising this uncertainty, and how policy-makers should cope with it, is the subject of the remaining sections.

6.3 A Changing Climate for Development

Many developing countries already face challenging climatic conditions; in general, they are hotter and experience more variable rainfall patterns than their richer counterparts (Stern 2007). It is widely anticipated that the effects of future climate change will exacerbate these climatic challenges in poorer parts of the world (IPCC 2013). The expectation that climate change will have its most damaging effects in poorer countries is based partly on projections of where future changes in climate will be most negative from a socioeconomic perspective (e.g. Samson et al. 2011), and partly on the observation that poorer countries are more vulnerable to changing climatic conditions, given their exposure (existing climate and reliance on agricultural output) and lack of adaptive capacity (e.g. Fankhauser and McDermott 2014).

The economic and development impacts of gradual changes in climate can be illustrated using historical data. For example, changes in moisture availability have been shown to have notable effects on agricultural productivity, rural–urban migration patterns and economic growth. Observed declines in moisture availability to date have been most pronounced in already arid areas, exacerbating existing vulnerabilities (Henderson et al. 2014).

Anthropogenic forcing is expected to result in global warming (i.e. increases in average temperatures) of anywhere between 2 °C and as much as 5–6 °C, by 2100, under different emissions scenarios. Such warming, however, will not be distributed evenly around the planet. Some locations will experience significantly more warming for any given global average change. For example, land warms faster than oceans. Similarly, there are well-understood physical mechanisms, which indicate that warming at high latitudes will be greater than at lower latitudes. Thus, for any given target for the increase in global mean temperatures (e.g. +2 °C), the implication is that most if not all land areas would warm by more than this, and higher latitude land areas by substantially more.

However, it is not just changes in mean temperatures or precipitation that matter for development. Climate change represents a change in the distribution of future weather (Daron and Stainforth 2013); investment decisions and economic activity more generally will be sensitive to more than the mean of that distribution (Stainforth et al. 2007a). From a DRM perspective, it is the frequency of extremes that is most relevant, since extreme temperatures, precipitation (both abundance and scarcity) and winds are generally the triggers of climatic disasters. Making projections about the future distribution of extreme weather events, at a scale relevant to policy-makers and investment decisions is even more challenging than predicting average changes (e.g. IPCC 2012). Again, however, we can make some qualitative predictions, based on physical principles.

Higher average temperatures would change the shape of local temperature distributions (e.g. Stainforth et al. 2013). A first-order expectation is that the shift towards higher average temperatures would involve more frequent extremes of heat and less frequent extremes of cold. Temperatures currently considered extremely hot may become the norm. The frequency of heavy rainfall events or the proportion of total rainfall from heavy falls is also expected to increase for many areas (IPCC 2012). This follows from the basic principle that a warmer atmosphere can hold more water. More intense rainfall episodes might lead to an increase in flood risk. The intensity of rainfall events might even increase in locations where total rainfall is anticipated to decline, leading to higher variability (ibid.). Drought intensity is also expected to increase, for some areas and seasons, although there is only medium confidence in these expectations (ibid.). For tropical cyclones, there is an expectation of fewer cyclones, but maximum wind speed is likely to increase, at least in some ocean basins—resulting in more destructive cyclones (ibid.). Higher global temperatures could also lead to increased flood risk via sea level rise, particularly threatening to low-lying, coastal or deltaic areas, which happen to include some of the most densely populated regions of the planet, many of them in developing countries (e.g. Bangladesh).

These qualitative assessments about what climate change will imply for future weather provide useful guidance for policy-makers in a fairly general sense. In particular, it is confidently expected that climate risk (both variability and extremes) will increase for many developing countries under most climate change scenarios. This expectation therefore reinforces the case for DRM investment, and highlights adaptation to climate change as an important potential co-benefit of DRM initiatives. Translating anticipated global changes into projections at the kind of temporal and spatial scales that are relevant for project evaluation-type decisions is considerably more challenging. The problem of uncertainty in climate change projections and the implications for DRM investment strategies are discussed in greater detail in the next sections. The focus here is on uncertainty related to climate change. However, it should be noted of course that DRM investments are also subject to uncertainty in relation to future exposure and vulnerability—both of which are at least partly dependent on development trajectories. As argued elsewhere in this chapter, DRM strategies also need to take account of the interaction of development trends with (evolving) climate risks.

Without anthropogenic forcing, weather is often considered to be chaotic; "under changing concentrations of atmospheric GHGs [greenhouse gases], the behaviour is not chaotic but pandemonium" (Stainforth et al. 2007a, p. 2147). Making (long-term, in some cases irreversible) investment decisions under climate change presents the challenge of dealing with deep uncertainty. Even a "perfect" climate model would produce a distribution of possible future weather trajectories, only one of which will ever be realised. If that distribution contains a large range of possible values for decision-relevant variables (e.g. precipitation quantity and timing, temperature averages and extremes, wind speeds, etc.), simply taking expectations (i.e. the mean) of that distribution, as would be common in economic analyses (or standard project evaluation techniques such as cost–benefit analysis), may be seriously misleading, particularly given the possibility of important thresholds or tipping points that may lie within the range of uncertainty (Kemp 2005).

Of course, there is no such thing as a "perfect" climate model. Uncertainties in relation to modelling future climate change derive from several distinct sources (Stainforth et al. 2007a); anthropogenic forcing, initial conditions and model imperfections (both model uncertainty and model inadequacy). These uncertainties are challenging enough for global climate models, which are relatively well understood. Global models can provide information relevant to mitigation decisions. However, adaptive and DRM investment decisions require climate information at a much finer spatial resolution. Attempts to "downscale" global climate projections to spatial and temporal scales relevant for adaptive investment decisions involve additional layers of uncertainty associated with "local physics, topography, and an incomplete understanding of how downscaling techniques interact with uncertainties already present in GCMs" (Millner 2012, p. 144). Uncertainty—specifically the disagreement between climate models—has been shown to increase as the spatial scale of climate projections is reduced (Masson and Knutti 2011).

One common approach to dealing with uncertainty over future climate change has been to produce a range of climate change projections, for example based on "ensembles" of different climate models, with varying initial conditions and forcings. The outputs of these various model runs are often combined into a single probability density function (PDF), by applying a weighting scheme (Tebaldi and Knutti 2007). This single PDF might be thought to characterise climate risk for a given variable, location, time period, etc., and the standard optimising techniques of economic analysis could then be applied to make investment decisions, taking account of risk–return trade-offs.

However, such weighted combinations of model outputs are likely to be highly misleading in that they imply a degree of confidence not justified by the underlying assumptions and the known inadequacy of our current models. In particular, Stainforth et al. (2007a, p. 2155) emphasise that "the lack of any ability to produce useful model weights, and to even define the space of possible models, rules out the possibility of producing meaningful PDFs for future climate based simply on combining the results from multi-model or perturbed physics ensembles". The inability to calculate probabilities with any confidence renders standard economic approaches to project evaluation (e.g. cost–benefit analysis) inadequate. Where DRM investments are sensitive to climate uncertainty, cost–benefit analyses need to be supplemented by additional screening devices, as discussed further in the next section.

Where standard economic analyses may demand precise probabilities, in reality any confident statements about future climate are likely to be of a qualitative nature, and reliant on a number of significant assumptions. However, Stainforth et al. (2007a) note that models can still provide insight, and even qualitative guidance can be valuable in informing adaptation decisions. In interpreting the output from climate models, these authors encourage users to view them as providing "a range of possibilities which need to be considered" (ibid., p. 2159). They characterise the best information that climate models or ensembles can currently provide as a "lower bound on the maximum range of uncertainty", or what they refer to as the "climate envelope" (ibid., p. 2155). They also stress that this range is "non-discountable" in the sense that "we should not disregard the possibility that the response could be anywhere within the envelope".

This last point is crucial for DRM investment decisions. From a policy- or decision-maker's perspective, the possibility that the outcome will be anywhere within that range cannot be disregarded. Standard evaluation techniques, such as cost–benefit analysis, where risks are discounted according to known or expected probability of their occurrence over a particular time period, might therefore produce very misleading policy recommendations. Stainforth et al. (2007b) instead advise using the boundaries of the "climate envelope" as an initial screening for adaptive investment decisions. For example, for a decision-relevant climate variable, one should consider the maximum and minimum projected values for that variable and whether either of these "extremes" might alter the planned investment. Would the most benign future climate scenario render the project unnecessary, or economically unviable in terms of avoided losses relative to costs? Would the "worst case" future climate scenario render the project inadequate, in terms of providing the desired level of protection?

More generally, risk assessments need to consider many combinations of values for key parameters, including those related to climate, exposure and vulnerabilities. The nature of uncertainty in relation to climate projections means the range of climate-related values in any such exercise needs to include the boundaries at either end of the climate envelope. We return in the next section to this idea of using the climate envelope as a screening device as part of a heuristic guide to DRM investments under uncertain climate change.

6.4 A Decision Framework for DRM Investment Under Climate Uncertainty

How should policy-makers treat the highly uncertain information available from climate models and associated analyses of socioeconomic impacts? One seemingly reasonable approach would be to plan for a central (most likely) scenario. However, such an approach is problematic, and risky, on a number of levels. For one thing, in a non-linear climate system, taking a central or mean expectation might under-represent the possibility of a threshold or tipping point being breached, resulting in some extreme or catastrophic scenario. One must also consider whether projects approved under a "most likely" scenario risk locking in development paths that would be vulnerable to more extreme scenarios (or, equally, whether such projects would ultimately prove to be unnecessary in the case that realised changes are less than anticipated). A classic example of this dilemma is flood defences. They must be built to some specification. But what is a reasonable—or indeed optimal—level of protection?

Deciding whether or not to invest in expensive infrastructure projects requires some form of project evaluation. The standard approach to project evaluation for investment decisions is to apply a cost–benefit analysis that compares the discounted expected value of future benefits (e.g. the value of avoided losses) with the anticipated costs of the investment. With well-defined and well-understood risks and uncertainties, optimum expected utility techniques demonstrably produce the best outcomes (Lempert and Collins 2007). However, under uncertain climate change, these conditions may not hold. Furthermore, it is now well established in the environmental economics literature that investments related to environmental problems generally have important additional characteristics that are neglected by these standard evaluation frameworks, in particular; uncertainty over future costs and benefits of the project; irreversibilities once the policy or investment has been approved; and the option of delaying action until more information becomes available (e.g. Pindyck 2002).

Alternative decision rules, for example based on the principles of minimising regrets (e.g. Heal and Millner 2014) or "robust" decision-making (e.g. Hallegatte 2009; Lempert and Collins 2007) offer the promise of formalised quantitative analysis under deep uncertainty. However, such techniques require substantial analytical or computational resources, for example in order to calculate expected costs and benefits under a potentially large numbers of "plausible" climate futures (Hallegatte et al. 2010; Lempert and Collins 2007). Given that decision-making capacity is itself a scarce resource in many developing countries, this section offers a simpler decision-making framework, incorporating Stainforth et al.'s idea of screening investment decisions against the boundaries (maximum and minimum) of the climate envelope.

The deep uncertainty over future climate change, and its physical and socioeconomic impacts, would appear to underscore the need for greater flexibility to be incorporated in the design of DRM strategies. While the need for flexibility in adapting to climate change is incontrovertible (e.g. Fankhauser et al. 1999), its relevance for DRM investments will be greater in some circumstances (locations and projects) than others, and in particular will depend on the climate sensitivity of the proposed investment and the range of possible climate futures (the breadth of the climate envelope) for decision-relevant variables. So for example, where screening the proposed investment against the climate envelope indicates that even the most extreme climate change projections (at either end of the spectrum) would not alter the investment decision, building in flexibility may be redundant, and would likely incur unnecessary additional costs.

Uncertainty therefore has qualitatively different implications for different types of DRM investment projects. Figure 6.2 and the following discussion provide a heuristic decision framework that enables policy-makers to determine the extent to which issues related to investing under uncertainty should influence DRM investment decisions.

In the absence of any uncertainty in relation to future climate, DRM investments should be guided by standard cost–benefit type analysis—comparing the expected benefits of the investment in terms of avoided losses with anticipated costs and the best alternative use of the funds (standard investment opportunity cost)—supplemented by due consideration of potential development and adaptation dividends and trade-offs, as discussed in Sect. 6.2.

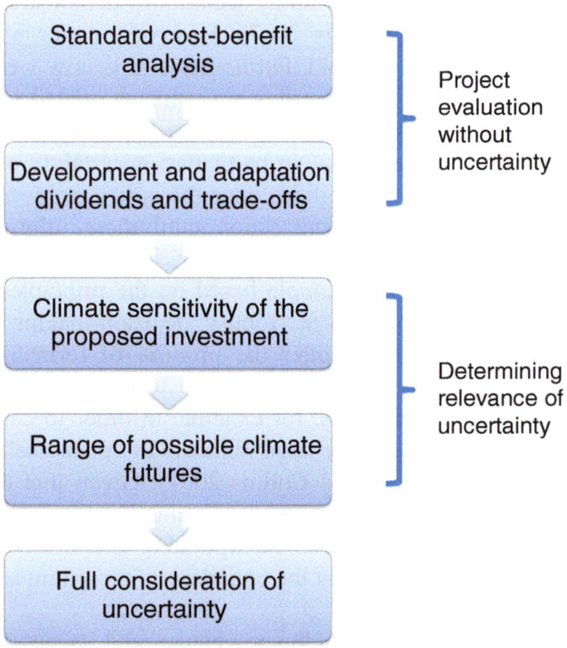

Fig. 6.2 Outline of decision framework for determining the role of uncertainty in project evaluation decisions

Under uncertain future climate change, the investment decision becomes more complex. However, this uncertainty is relevant only under certain conditions. Figure 6.3 provides a full illustration of how climate uncertainty can be incorporated into a project evaluation framework. First, we should consider if the investment itself is sensitive to changes in climate risk. In other words, are the expected benefits, in terms of avoided losses, dependent on the severity of some climate-related hazard? Flood defences and zoning or planning restrictions are clearly highly climate-sensitive in this sense. On the other hand, some DRM initiatives, such as disaster preparedness, emergency planning procedures and various risk-coping strategies, are likely to be less sensitive to changing climate risk and therefore require less detailed consideration of climate uncertainty before they can be safely adopted or rejected (postponed).

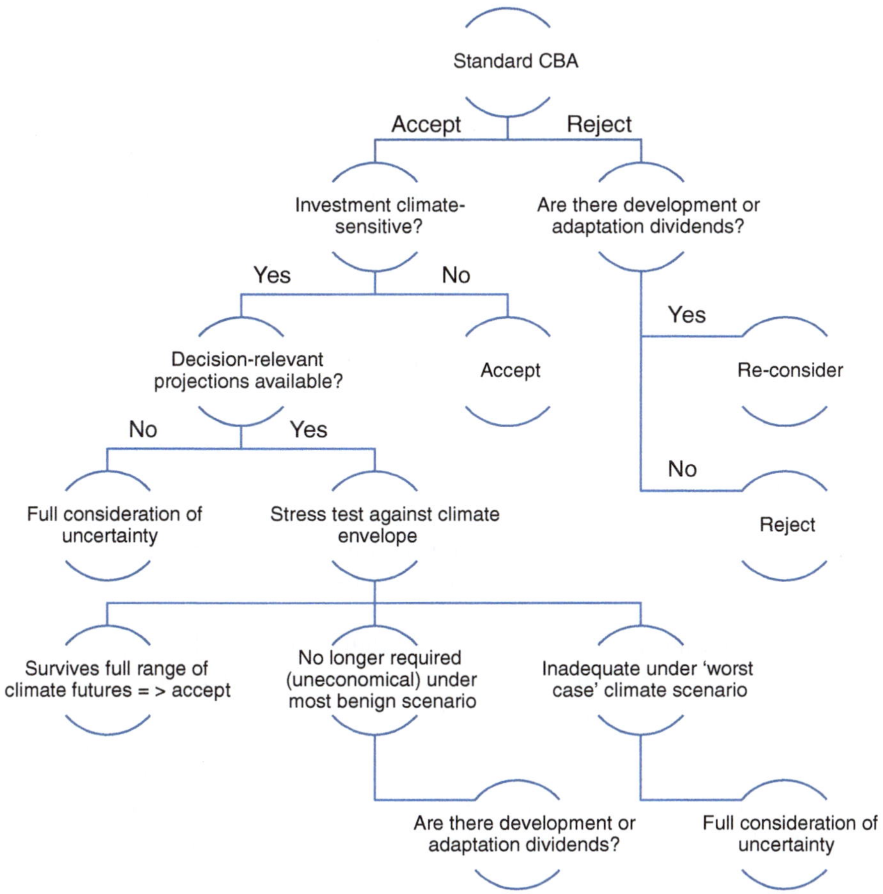

Fig. 6.3 Full "decision tree" for project evaluation under climate uncertainty

If the proposed project is climate-sensitive, and assuming projections for decision-relevant variables are available at the appropriate temporal and spatial scales, the project should be screened against the full range of possible climate futures. Such a screening process in practice involves just two additional calculations—one at each of the boundaries (maximum and minimum) of the climate envelope for the decision-relevant variable. If the project survives the full range of the climate envelope then it can be accepted without further consideration of issues related to uncertainty.

Rejection of the project at either boundary of the climate envelope will have distinct implications for that investment decision. If under the most benign climate future the project is no longer required (i.e. becomes uneconomical), then the risk of redundancy needs to be considered and alternative uses of scarce resources may be preferred. However, such a project may still be worth pursuing, particularly where there are (large) anticipated development or adaptation dividends (or other social, environmental or economic co-benefits—see Chap. 1).

On the other hand, if under a "worst case" climate scenario the project would be rendered inadequate in terms of the level of protection provided, a more complete consideration of uncertainty and how it relates to the specific characteristics of the proposed investment would be required. This would also apply to the situation of a climate-sensitive investment project where no decision-relevant projections are available.

A first-order consideration for projects that fail the worst case climate scenario stress test (or where decision-relevant projections are unavailable) would be the risk of locking in further vulnerability, such as in the case of flood defences, where the concentration of population and economic assets exposed might be increased as a result of the protection provided (moral hazard risk). In other words, could the proposed investment actually make things worse under some climate futures? If so, the project is in danger of exacerbating (rather than mitigating) disaster risk. Such scenarios are obviously highly undesirable and would therefore suggest a rejection of the proposal.

If, on the other hand, these (moral hazard) risks are not present, the project could be allowed to proceed, provided it meets two further criteria. First, could the project be scaled in response to evolving risk (or risk evaluations)? In other words, is there flexibility in the design to increase or decrease the level of protection over time? Second, are there partial benefits from different degrees of intervention, or does the project need to achieve a certain scale (or be fully completed) before any benefits are conferred (time-to-build concerns—see e.g. Millner 2012). If the proposed project has both partial benefits (i.e. doesn't face time-to-build concerns) and flexibility in design (i.e. the level of protection is scalable in response to changing risk profiles) then investment might still be justified, even where the project has failed the worst case climate scenario stress test. If these characteristics are not (both) present, however, this suggests the project should be rejected (or postponed, pending new information).

6.5 The Optimal Timing of Investment and the Value of Waiting for New Information

The literature on investing under uncertainty emphasises that, where investments are at least partly "irreversible", there is an opportunity cost to investing today in the form of the foregone opportunity to wait and learn from new information (e.g. Dixit and Pindyck 1994; Pindyck 2002). The investment decision is then not just whether or not to invest, but also whether it is optimal to invest today or wait for new information. The potential value of waiting for new information will depend on the degree of irreversibility of the investment and the associated value of avoided "regret" in the case of its adoption.

Uncertainty does not automatically imply a "wait and see" approach to DRM. In the first instance, the value of waiting for new information (in terms of avoided regret) must be weighed against the likelihood of its timely arrival, which is not guaranteed in the case of climate prediction.

Under climate change, the prospects for significant improvements in our ability to make reliable forecasts at decision-relevant scales appear pretty dim (Heal and Millner 2014; Stainforth et al. 2007a). The various sources of uncertainty in climate projections become relevant at different time scales and also differ in the extent to which we can expect improvements in our understanding and predictive capacity (as discussed in Millner 2012). Over shorter time horizons, total prediction uncertainty is dominated by internal variability (initial conditions) and model imperfections, with internal variability increasingly important at smaller spatial scales and on shorter time horizons. The latter, in particular, may be amenable to gradual improvements through a better understanding of initial conditions, based on improvements in observations (Smith et al. 2007). The uncertainty over longer-term projections, on the other hand, is dominated by uncertainty over future emissions, which is essentially unknowable (Millner 2012).

Adaptive investments with high adjustment costs (i.e. those that involve longer-term commitments, less flexibility or elements of irreversibility) require the greatest precision in (longer-term) forecasts, before one can safely (unreservedly) recommend adoption or rejection (postponement) of that investment (Millner 2012). Under uncertain climate change, this finding would suggest DRM investments favour more flexible initiatives over those with large sunk costs and elements of irreversibility. However, this analysis does not suggest hard defensive infrastructure investments should never be undertaken, but rather that the uncertainty over future climate change increases the risks associated with such projects and therefore places a greater burden of proof on their advocates to ensure the benefits in the absence of climate change are sufficiently large, and unlikely to be reversed under a large range of possible climate futures.

There are two kinds of irreversibility and associated potential for regret related to uncertain DRM investment decisions (Pindyck 2002). These work in opposite directions. One is the sunk cost associated with an investment—an obvious example is the case of hard infrastructure projects (once built they are essentially fixed), but sunk costs may also be in the form of political constraints making a DRM policy (e.g. land zoning) difficult to reverse once implemented. These sunk costs make adopting a policy today more risky than would be implied by a standard cost–benefit analysis.

The second type of irreversibility is a "sunk benefit", or negative opportunity cost, of adopting the policy now rather than waiting; this relates to foregone adaptive opportunities (Fankhauser et al. 1999), which are missed while one waits for new information. Such adaptive opportunities might be most strongly associated with initiatives that attempt to guide development trends towards more sustainable trajectories—as discussed earlier. These sunk benefits (or adaptation dividends) strengthen the case for adopting a policy today, relative to what would be implied by a standard cost–benefit analysis. DRM policies that reduce future exposure to climate risk convey a future "development dividend" by improving the resilience of development gains.

Even if we were to adopt a "safety first" policy (precautionary principle), this would still not necessarily favour a "wait and see" approach to DRM policy. Inaction (postponement of the investment) is also an active policy choice (or at least should be treated as such), and one that carries its own set of risks and potential for regret. These are most pronounced in situations that involve a degree of irreversibility in the investment decision.

Where the irreversibility is on the side of the proposed investment, potentially locking in exposure to future risk (sunk cost and moral hazard risk)—for example a flood defence system that carries moral hazard risk in the event of its failure—uncertainty implies that adoption of the proposal may not be safe. In other words, a safety first approach in this case would favour postponement of the investment. On the other hand, where the irreversibility is on the side of the proposed investment, potentially avoiding locking in future risk (sunk benefit or adaptation dividend)—for example land use or zoning restrictions that attempt to guide existing development trends to avoid creating long-term, irreversible exposure to climate hazards—uncertainty implies it may not be safe to postpone the investment. In other words, a safety first approach in this case would favour early adoption of the proposed DRM initiative.

There are relatively limited circumstances in which a full consideration of uncertainty and how it interacts with the specific characteristics of a proposed DRM investment project would be required. Specifically, this is only the case for projects that are in the first instance (highly) climate-sensitive and additionally where either (1) no decision-relevant projections are available or (2) under the worst case climate scenario the proposed project would be rendered inadequate in terms of the level of protection provided. Where climate uncertainty is most relevant, it would appear to shift the balance of appropriate investment strategies away from those with a large component of commitment or irreversibility, towards more flexible, low-regret interventions. However, as noted in this section, a safety first approach is not equivalent to advocating a "wait and see" attitude to DRM investments. In some cases, particularly where DRM could help avoid locking in future exposure to climate risk, uncertainty provides additional motivation for the early adoption of DRM policies.

6.6 Conclusions

The expectation that climate risk (both variability and extremes) will increase for many developing countries under climate change reinforces the case for DRM investments and highlights adaptation to climate change as an important potential co-benefit of DRM initiatives.

Uncertainty over the precise climate risk that will be faced at any given location represents a "known unknown" for DRM strategies. The nature of climate projections is such that they offer a very uncertain picture of the future at the kind of spatial and temporal scales that are relevant for project evaluation decisions. However, ignoring uncertain future climate risk could result in exposing DRM investments to large costs in the form of maladaptation and missed opportunities for adaptation or development dividends.

While the deep uncertainty associated with climate change complicates DRM investment decisions, the analysis presented here shows these considerations are relevant only for a relatively limited set of investment circumstances. This chapter offers a decision-making framework that attempts to simplify the process of accounting for the deep uncertainty associated with climate projections and the specific characteristics of different DRM policy options. In particular, this framework enables policy-makers to identify the particular circumstances under which uncertainty about future climate change becomes critical for DRM investment decisions. It also emphasises two important elements of successful DRM strategies; the first is to give careful consideration to alternative uses of scarce resources—which is particularly crucial in a development context—and potential development or adaptation co-benefits. The second is to stress test DRM policies against the boundaries of the climate envelope—in other words to consider not just "likely" or expected scenarios but also if the proposed intervention remains worthwhile even under the most benign or worst case projected climate outcomes.

Uncertainty related to future climate change does not necessarily motivate a "wait and see" approach to DRM investments. Instead, the analysis here has demonstrated that, where opportunities exist to avail of adaptation dividends—for example where DRM initiatives could help avoid locking in future exposure to climate risk, climate uncertainty provides additional motivation for early investment in DRM initiatives.

An optimistic message also emerges from this analysis, which has identified substantial overlap between the flexible, low-regret-type interventions favoured under uncertain climate change and the risk-coping initiatives that are likely to have the greatest co-benefits for economic development. Such policies deliver economic and social benefits (development dividends), regardless of the climate future that materialises in a given location. These flexible, low-regret-type DRM policies are also less likely to incur moral hazard risk, which may be associated with hard defensive infrastructure investments such as flood defences. This chapter therefore presents a case for greater investment in DRM initiatives, conditional on these being designed with an explicit development-first approach and due consideration of uncertainty over future climatic conditions. In short, appropriate DRM strategies should focus on supporting development paths that are robust to a range of possible climate futures. Such strategies would have the dual benefits of maximising potential co-benefits of DRM investments for development, while minimising regret under uncertain climate change.

References

Bryan G, Chowdhury S, Mobarak AM (2014) Underinvestment in a profitable technology: the case of seasonal migration in Bangladesh. Econometrica 82(5):1671–1748

Collier P, Conway G, Venables T (2008) Climate change and Africa. Oxford Rev Econ Policy 24 (2):337–353

Daron J, Stainforth D (2013) On predicting climate under climate change. Environ Res Lett 8:1–8

Davis DR, Weinstein DE (2002) Bones, bombs, and break points: the geography of economic activity. Am Econ Rev 92(5):1269–1289

Dercon S (2012) Is green growth good for the poor? Policy Research Working Paper 6231. World Bank, Washington, DC

Dixit A, Pindyck R (1994) Investment under uncertainty. Princeton University Press, Princeton, NJ

Fankhauser S, McDermott TKJ (2014) Understanding the adaptation deficit: why are poor countries more vulnerable to climate events than rich countries? Glob Environ Change 27:9–18

Fankhauser S, Smith J, Tol R (1999) Weathering climate change: some simple rules to guide adaptation decisions. Ecol Econ 30(1):67–78

GFDRR (Global Facility for Disaster Risk Reduction) (2015) Investing in resilience. World Conference on Disaster Risk Reduction, Sendai, 18 March

Hallegatte S (2009) Strategies to adapt to an uncertain climate change. Glob Environ Change 19:240–247

Hallegatte S (2013). A normative exploration of the link between development, economic growth, and natural risk. Mimeo

Hallegatte S (2014). Economic resilience. Policy Research Working Paper 6852. World Bank, Washington, DC

Hallegatte S, Henriet F, Patwardhan A et al (2010) Flood risks, climate change impacts and adaptation benefits in mumbai: an initial assessment of socio-economic consequences of present and climate change induced flood risks and of possible adaptation options. Environment Working Paper 27. OECD, Paris

Heal G, Millner A (2014) Uncertainty and decision making in climate change economics. Rev Environ Econ Policy 8(1):120–137

Henderson JV, Storeygard A, Deichmann U (2014) 50 years of urbanization in Africa. Policy Research Working Paper 6925. World Bank, Washington, DC

IPCC (Intergovernmental Panel on Climate Change) (2012) Special report of the IPCC: managing the risks of extreme events and disasters to advance climate change adaptation. Cambridge University Press, Cambridge

IPCC (Intergovernmental Panel on Climate Change) (2013) Climate change 2013: the physical science basis. Contribution of Working Group 1 to the Fifth Assessment Report of the IPCC. Cambridge University Press, Cambridge

Kellett J, Caravani A (2014) Financing disaster risk reduction: a 20 year story of international aid. Overseas Development Institute, London

Kemp M (2005) Science in culture: inventing an icon. Nature 437(1238). doi:10.1038/4371238a

Lempert R, Collins M (2007) Managing the risk of uncertain threshold responses: comparison of robust, optimum, and precautionary approaches. Risk Anal 27(4):1009–1026

Masson D, Knutti R (2011) Climate model genealogy. Geophys Res Lett 38:L08703. doi:10.1029/2011GL046864

Mechler R, Bouwer LM, Linnerooth-Bayer J, Hochrainer-Stigler S, Aerts J, Surminski S, Williges K (2014) Managing unnatural disaster risk from climate extremes. Nat Clim Change 4:235–237

Millner A (2012) Climate prediction for adaptation: who needs what? Clim Change 110:143–167

Mobarak A, Rosenzweig M (2013) Informal risk sharing, index insurance, and risk taking in developing countries. Am Econ Rev 103(3):375–380

Oreskes N, Stainforth D, Smith L (2010) Adaptation to global warming: do climate models tell us what we need to know? Philos Sci 77:1012–1028

Pindyck R (2002) Optimal timing problems in environmental economics. J Econ Dyn Control 26:1677–1697

Samson J, Berteaux D, McGill B, Humphries MM (2011) Geographic disparities and moral hazards in the predicted impacts of climate change on human populations. Glob Ecol Biogeogr 20(4):532–544

Smith D, Cusack S, Colman A, Folland C, Harris G, Murphy J (2007) Improved surface temperature prediction for the coming decade from a global climate model. Science 317 (5839):796–799

Stainforth D, Allen M, Tredger E, Smith L (2007a) Confidence, uncertainty and decision-support relevance in climate predictions. Philos Trans Roy Soc A Math Phys Eng Sci 365(1857):2145–2161

Stainforth D, Chapman S, Watkins N (2013) Mapping climate change in European temperature distributions. Environ Res Lett 8(3). doi:10.1088/1748-9326/8/3/034031

Stainforth D, Downing T, Washington R, Lopez A (2007b) Issues in the interpretation of climate model ensembles to inform decisions. Philos Trans Roy Soc A Math, Phys Eng Sci 365 (1857):2163–2177

Stern NH (2007) The economics of climate change: the stern review. Cambridge University Press, Cambridge

Tanner T, Surminski S, Wilkinson E, Reid R, Rentschler J, Rajput S (2015) The triple dividend of resilience: realising development goals through the multiple benefits of disaster risk management. Global Facility for Disaster Reduction and Recovery (GFDRR) at the World Bank and Overseas Development Institute (ODI), London

Tebaldi C, Knutti R (2007) The use of the multi-model ensemble in probabilistic climate projections. Philos Trans Roy Soc A Math Phys Eng Sci 365(1857):2053–2075

Chapter 7
Financial Crises and Economic Resilience: Lessons for Disaster Risk Management and Resilience Dividends

Stephany Griffith-Jones and Thomas Tanner

Abstract The development progress achieved by many countries, and particularly by low-income countries, is at risk of being undermined or even wiped out by the range of shocks and resulting crises they face. Since the turn of the millennium, there has been growing recognition of the importance of climate and disaster risks for development progress; the global financial crisis of 2007/08 also had profound implications for economies around the world. Partly in response to this experience, anticipatory risk management systems have become an increasingly popular approach to tackling both economic and disaster resilience. This chapter examines the impacts of financial crises on development at the national level and the responses of major international institutions in terms of coping with and anticipating such shocks. It then examines the lessons from these risk management mechanisms for understanding and recognising the dividends of resilience emerging from disaster risk management.

Keywords Disaster risk · Financial crisis · Economic resilience · Co-benefits

7.1 Introduction

The development progress achieved by many countries, and particularly by low-income countries (LICs), is at risk of being undermined or even wiped out by the range of shocks and resulting crises they face. Since the turn of the millennium,

Many thanks to Swenja Surminski, Mook Bangalore and anonymous reviewers for insightful comments and inputs.

S. Griffith-Jones (✉)
Columbia University, New York, USA
e-mail: sgj2108@columbia.edu

T. Tanner
Overseas Development Institute, London, UK
e-mail: t.tanner@odi.org.uk

S. Surminski and T. Tanner (eds.), *Realising the 'Triple Dividend of Resilience'*, Climate Risk Management, Policy and Governance,
DOI 10.1007/978-3-319-40694-7_7

there has been growing recognition of the importance of climate and disaster risks for development progress. The global financial crisis of 2007/08 also had profound implications for economies around the world (Benson and Clay 2003; Te Velde et al. 2011). In response, anticipatory risk management systems have become an increasingly popular approach to tackling both economic and disaster resilience. This chapter examines the impacts and costs of financial crises on development at the national level and the responses of major international institutions in terms of coping with and anticipating such shocks. It then explores the lessons from these financial risk management mechanisms for promoting investment in resilience by emphasising the co-benefits of investment that can be realised even in the absence of future shocks.

Crucially, while there has been a more concerted effort and investment to tackle uncertainty and damage caused by financial crises, as well as shocks through trade channels, investment and progress in disaster risk management (DRM) has regularly been outstripped by the increase in hazard burdens and vulnerability around the world (UNISDR 2015). There are a range of reasons for this underinvestment, including limited understanding of risks and impacts, a sheer lack of resources in poor countries, political myopia and attractiveness of more visible post-disaster support initiatives, as well as pressure to use scarce resources to respond to other urgent competing needs such as infrastructure, education and health (Vorhies 2012; Wilkinson 2012; World Bank 2013).

Beyond these relatively well-known challenges there is one aspect that has only recently gained the attention of DRM and development experts. There is growing evidence that underinvestment also occurs owing to a failure to capture the wider range of development dividends that DRM creates, both through reductions in the background level of risk that enable individuals to take positive risks (e.g. innovation and entrepreneurship) and through the 'ripple effects' of wider social, environment or economic co-benefits of investment (Rodin 2014; Tanner et al. 2015). In other words investment in DRM makes sense even in the absence of disasters, which presents a strong argument in favour of undertaking these preventive measures. Financial crisis prevention and management mechanisms can similarly be defended for the wider benefits of stability and growth, including for the overall global system.

The impacts of the series of financial crises hitting developing countries in the past three decades have been compounded by the effects of the global financial crisis of 2007/08, which started in the industrialised economies. As a result, growing concern has arisen about uncertainty from the world economy and particularly about the lower resilience of developing and emerging economies to external shocks, which undermine their long-term development (Didier et al. 2012; Griffith-Jones and Ocampo 2009). This concern has presented itself both among industrialised and industrialising countries that are increasingly integrated into global trading and financial systems and among countries that are the poorest, smallest and most vulnerable to external shocks. The latter group have faced significant challenges to their financial capacity to address their vulnerabilities, rebuild their pre-2007/08 crisis financial buffers and build new capacities for resilience in the face of persistent or frequent crisis. The situation has also led to uncertainty about how resources for development will be secured, on the scale and with the degree of reliability needed to absorb new external shocks as they come, so enabling these countries to eradicate poverty and to achieve higher levels of growth and sustainable development.

Experience during both recent and earlier financial crises has led to growing consensus that external shocks can disrupt both short-term growth and long-term development. There is clear evidence that financial crises have become more frequent and more damaging, as economies and financial systems have become more integrated within the global economy, and as financial systems have become more liberalised, without corresponding regulation (Griffith-Jones and Gottschalk 2012). The response to exogenous economic shocks (and especially shocks arising from financial crises) has been twofold: (1) crisis management, including provision of international liquidity and development finance to countries hit by shocks, to help sustain both short-term growth and long-term development, and (2) crisis prevention, to help make crises less likely and smaller if they do occur.

Strengthening the resilience of financial systems can yield benefits in the event of financial crises, but also provide greater stability outside times of crisis. This chapter therefore looks to draw lessons from growth in financial risk management mechanisms for the policy and practice of DRM. Section 7.2 summarises the impact of economic shocks on national and international economies in terms of short- and long-run growth and development. Section 7.3 then examines international responses to the global economic crisis of 2007/08, with analysis of the International Monetary Fund (IMF) and World Bank responses in Sect. 7.4. Sections 7.5 and 7.6 draw lessons for DRM from understanding the mechanisms for improved financial risk management at the national and international level.

7.2 The Case for Prevention: Financial Crises and the Costs to Growth and Development

7.2.1 *Costs of Crisis to Economic Growth*

Addressing shocks once they occur, or minimising the likelihood of their occurrence by preventive action, is vital to achieving long-term growth and development. External economic shocks tend to have very large negative effects on developing economies' growth, investment and poverty. When a developing country suffers an external shock, the balance of payments, the fiscal accounts and the overall level of economic activity suffer. The initial effects on these key macroeconomic variables feed through the entire economy, with very negative social and economic effects taking place through reduced tax revenue, lower government spending, lower private and public investment, lower wages, higher unemployment and therefore higher poverty.

External economic shocks traditionally came more through the trade channel, as developing economies were integrated into the global economy through trade. Economies therefore suffered in the event of declines in the prices of one or more of the main exports of a country, a fall in their volume or an increase in the price of imports. Such shocks could be temporary or more permanent. If the shock were temporary (e.g. brief deterioration of terms of trade) and were to be financed quickly for a high proportion of the shock through official liquidity, any negative impact on growth and poverty could be avoided. Official liquidity could allow levels of imports to be maintained, which implied maintaining economic activity. This was what the IMF created the Compensatory Financing Facility (CFF)—the first official multilateral liquidity facility created for this purpose by the international community—for in 1963 (Griffith-Jones 1983).

More recently, and as a result of increased integration of a growing number of economies into private financial flows, shocks have more frequently come from the capital account, because of either changes in the level of net capital flows and/or their cost. Often, such changes in the net capital flows or their cost have implied a very high proportion of a country's gross domestic product (GDP). Countries hit by the 1997/98 East Asian crisis experienced a reversal of net capital flows of more than 10 % of GDP, leading to currency and banking crises, with significant costs to growth and investment. For some countries, changes in the level of remittances also became a potential source of external shocks, especially during the global financial crisis, which originated in the developed economies and started in 2007/08. In this and other crises, several exogenous economic shocks occurred simultaneously, hitting developing and emerging economies through different channels at the same time.

Evidence shows the problem in poor countries is not just a failure to achieve long periods of sustained economic growth but also the frequency of downturns (Winters et al. 2010). Low-income countries (LICs) increased their per capita GDP by only 11 % between 1960 and 2007. Either halving negative growth rates, by halving the severity of downturns, or halving the percentage of years of negative growth between 1960 and 2007 would have increased GDP by about 70 %. But if negative growth rates could have been completely eliminated, GDP per capita for LICs would have more than doubled, with average annual growth increasing significantly to over 2 % from the 0.23 % achieved over this period (ibid.).

Poor countries remained poor because they have periods of deeply negative growth that more than cancel out prior periods of positive growth. Such periods of negative or low growth are often caused by external shocks. LICs are often poorly equipped to deal with, and recover from, adverse shocks (Aiello 2009). Consequently, there is growing consensus that international shocks financing is particularly significant for low-income and small vulnerable economies, especially as they become more integrated into the world economy. This has parallels with climate change-related and other disasters, with poorer countries generally more at risk from impacts at the same time as having fewer resources with which to invest in DRM and climate adaptation (Olsson et al. 2014). Combined with ethical issues around the unequal distribution of causes of climate change, this has reinforced calls for greater external resources for these purposes.

Financial crises have very high costs for emerging and industrialised economies. Eichengreen (2004) estimated the cost of currency and banking crises at 0.7 % of developing country/emerging market GDP per year, equivalent to an annual amount of $107 billion. His estimates draw both on historical work that estimates output losses by examining crises during the past 120 years and on looking at average output losses per year during the 1980s and 1990s in Latin America and Asia. Eichengreen estimates that, during 25 years, currency and banking crises reduced incomes of developing countries and emerging economies by around 25 %. Griffith-Jones and Gottschalk (2007) estimated the output loss emerging market countries suffered between 1995 and 2002, when crises were prevalent in emerging economies, as a direct result of major currency and twin crises, by comparing potential and real economic output. They estimated an annual average of around $150 billion of lost GDP for that period, implying a total loss of $1250 billion for the 1995–2002 period (a figure similar to but somewhat higher than that of Eichengreen). The forgone output in that period resulting from crises corresponds to 54 % of the combined GDP of the East Asia and Pacific region and 65 % of the combined GDP of Latin America and the Caribbean in one year (Griffith-Jones and Gottschalk 2007).

Looking at a very large number of financial crises, Reinhart and Rogoff (2008) estimate that banking crises lead to an increase of the unemployment rate of on average 7 %, lasting for on average four years. The recent Eurozone crisis has led to even far higher increases in unemployment, with explosive growth in unemployment among the young, especially in Greece and Spain. Reinhart and Rogoff further estimate output falls (from peak to trough, so not considering output lost as economy grew less than trend) of an average of over 9 %. Some financial crises have led to far higher declines in output. One recent example is that Greek GDP has fallen by over 25 % since the financial crisis started there.

A US Federal Reserve Bank of Dallas paper (Atkinson et al. 2013) estimates the cost for the US economy of the recent US financial crisis as an output loss of between $6 and $14 trillion between 2008 and mid-2013. The paper compares output to a baseline trend that might have existed absent the crisis, arguing that this amounts to the equivalent of 40–90 % of one year's economic output of the US. Per US household, the cost is estimated at $50,000–$120,000. It is noteworthy that studies highlight that total costs may be higher if long-term growth does not return to pre-crisis levels. They also stress other dimensions, such as lower employment and individual welfare.

One important area that is important relates to the negative effects that interruptions to growth caused by exogenous economic shocks or climate shocks have on private investment. Investment, especially lumpy infrastructure, often declines during a crisis because investment decisions are sensitive to uncertainty about the future outcomes of key variables (Dixit and Pindyck 1994). An increase in uncertainty can change the investment decision and lead to the cancellation or at least postponement of lumpy investment projects with long-term negative implications for development. In key social aspects (such as reduced nutrition and withdrawal of children from school), countries that recover growth after a period of no or negative growth can see irreversible costs for their economies and for the poor (Cornia et al. 1987; Harper et al. 2009). Severely malnourished children or those who have missed longer periods of schooling may suffer effects for the rest of their lives and can even pass these negative effects to their own children, implying greater future poverty and lower prospects for growth.

7.2.2 The Growth in Preventative Action

Developing countries are increasingly recognising their inherent structural vulnerability to exogenous shocks, including both economic shocks and those from the natural environment, and have highlighted the need to review, systematise and expand shocks facilities (Griffith-Jones and Gottschalk 2012). The international financial institutions (IFIs), such as the IMF, the World Bank and the regional development banks (RDBs), as well as regional bodies like the European Commission (EC), have increasingly (and especially since the 2007/08 financial crisis) accepted the importance of shocks facilities and have moved forward to expand existing ones, as well as creating new ones, as we detail below.

As financial crises become more frequent, deeper and thus more costly, and spread more widely via contagion owing to financial globalisation, there is growing acceptance by institutions like the IMF and the World Bank of the necessity to devote an increasing share of their resources to fund developing and emerging countries' needs arising from external economic shocks. There is growing consensus that official international liquidity and development finance (both concessional and non-concessional), as well as grants, need to play an important role in mitigating the impact of economic exogenous shocks. This seems to clearly imply the desirability of allocating a higher proportion of official resources to shock financing in order to help developing and emerging countries, especially those that are more vulnerable and lack resilience to address shocks.

An important policy question beginning to be discussed as regards to economic shocks is therefore whether more emphasis should be placed on dealing with shocks, to help avoid growth declining in the short term and therefore harming long-term development and poverty alleviation, and, more specifically, how the potential trade-off in allocating less funds to other development activities can be addressed, as well as minimised. One way to reduce such a trade-off is to use the resources to increase resilience, especially if funds allocated to shocks for a certain period are not used during most of the period. Another way is to emphasise the overall benefits that can be derived from an anticipatory risk management approach, creating stability and favourable institutional conditions irrespective of whether or not crises occur.

7.3 Enhanced International Support Following the 2007/08 International Financial Crisis

Since the 1960s, the IFIs and the EC have put in place a range of so-called compensatory or shocks facilities to help countries deal with the above-described shocks, mainly focusing on economic shocks but increasingly including natural hazards. However, the global financial crisis that began in 2007/08 prompted IFIs to make a significant effort to attempt to shelter developing and emerging economies from the resulting shock, with the aim of protecting their growth and poverty reduction. They did this by both increased lending through existing facilities and by creating new facilities, as well as expanding the limits on existing ones. The increase in the capital of the World Bank and the RDBs, as well as in the resources available to the IMF, were important to facilitate the granting of significantly more credit.

At a conceptual level, there seemed to be a significant breakthrough in terms of recognising the important counter-cyclical function, which development banks (such as the World Bank and the RDBs) had to play in light of major events like the global financial crisis, especially to help sustain priority investment (both public and private). This investment had often been cut in the past when exogenous economic shocks hit countries, damaging future development. Thus, it was important not just to have additional international official liquidity as typically provided by the IMF to deal with Balance of Payments aspects (although this was clearly key where countries became foreign exchange constrained), but also to have counter-cyclical official international development finance, via both concessional and non-concessional lending, as well as increased grants where appropriate.

In what follows, we describe the main features of this international response, which was on a large scale, albeit significantly smaller than the contraction of private flows.

The IFIs—including the IMF, the World Bank and the RDBs—increased their lending to developing countries very significantly as a response to the global financial crisis. This had a positive impact in terms of ameliorating negative effects from the financial crisis on these countries' growth, investment and poverty reduction. Total lending commitments to developing countries jumped dramatically, from around $50 billion in 2007 to around $175 billion in 2009 and thereafter to an average of just over $200 billion annually in 2010–2011, resulting in a quadrupling of total lending commitments between 2007 and 2010–2011 (Griffith-Jones and Gottschalk 2012). Particularly large was the increase in IMF lending during those years, but the World Bank and RDBs also increased their lending significantly.

This response was significant and covered a large proportion of African, Caribbean and Pacific (ACP) countries' export shortfalls, especially in 2007 and 2008 (Te Velde et al. 2011). This large compensatory and generally counter-cyclical IFI lending, combined with other domestic factors and measures, not only helped avoid crises but also limited growth declines in developing countries in the face of major external shocks. These domestic factors included the existence of valuable buffers in developing countries, including high fiscal space (linked to prudent fiscal policies in good times), in many cases high levels of foreign exchange reserves and lower levels of external debt, as well as fairly prudent domestic financial regulation. Indeed, developing countries accumulated higher levels of foreign exchange reserves and lower levels of external debt, as well as regulating their domestic financial systems better as protection against future financial crises, based on their previous experience of financial crises, which had been so costly in terms of their development. Here, an important parallel, or even lessons, for natural disaster management can be drawn: taking preventive action at a national level *ex-ante* may have valuable pay-offs later, once an exogenous economic shock or a natural disaster hits.

In the event, growth in LICs, which had averaged 6.5 % annually in 2005–2007, fell to 5.7 % in 2008 and to 4.7 % in 2009, according to World Bank data. Although undesirable, such a decline could have been far worse given the magnitude of the shocks (especially in these cases on the trade account), and was followed by recovery to an average of 6.0 % growth in 2010–2011.

The global financial crisis hit middle-income country (MIC) growth more seriously, largely because MICs are more closely integrated with the international economy, especially via private capital flows; as discussed below, the contraction in private capital flows to these countries was initially so large that official flows—even though significant—could compensate for this decline only very partially MIC growth, which reached 8.0 % annually in 2005–2007, fell to 5.7 % in 2008, and significantly to only 2.6 % in 2009, although it recovered to 7.0 % annually in 2010–2011.

There are two important caveats to this overall fairly impressive response to the 2007/08 crisis by the IFIs. Perhaps most importantly, total lending commitments to LICs went up by far less than the total for all developing countries, from $17.5 billion annually in 2007–2008 to over $23 billion annually in 2009–2011—that is, by around 33 %, significantly less than the increase in commitments to MICs, which grew from $34 billion annually in 2007–2008 to $179 billion annually in 2010–2011, a rise of over 430 % (see Table 7.1).

Secondly, the most important increases only happened in 2009–2010, well after the crisis started. This picture is even clearer if we look at actual disbursements, which often lagged commitments quite significantly. Although International Development Association (IDA) commitments increased quickly in 2007 and 2009, disbursements hardly grew in those years, and they increased only modestly in 2008

Table 7.1 LIC and MIC lending commitments for all IFIs ($ millions)

	2006	2007	2008	2009	2010	2011
LIC	14,516	17,582	17,416	23,092	23,456	23,630
MIC	40,234	32,092	36,584	149,266	179,594	178,937
Total	54,750	49,674	54,000	172,358	203,050	202,568

Source Griffith-Jones and Gottschalk (2012)

and especially in 2010—that is, well after the shocks had hit LICs. There were also delays for World Bank [International Bank for Reconstruction and Development (IBRD)] lending to MICs. IFIs' cumulative disbursements are continually less than commitments from 2008 onwards, with the cumulative "disbursements gap" increasing to over $250 billion by 2011 (Griffith-Jones and Gottschalk 2012).

7.4 International Financial Crisis: Responses of the IMF and the World Bank

In what follows, we examine and evaluate recent reforms made to shocks facilities at the IMF (both concessional and non-concessional) and the World Bank (focused on IDA).

7.4.1 A Reform to IMF Shocks Facilities for LICs and MICs

In recent years, the IMF has made important changes to its lending facilities, especially as a response to the global financial crisis. Above all, it responded rapidly to the crisis, with commitments to developing and emerging countries increasing very sharply from $2 billion in 2008 to $83 billion in 2009 and $127 billion in 2011. This included, for Poverty Reduction Growth Facility-eligible countries (basically the LICs) increases in commitments from $657 million in 2008 to $1.5 billion in 2009 and $3 billion in 2010. The fairly strong increase of IMF lending to LICs was facilitated by a doubling of access as a percentage of quotas for all facilities in 2009.

Although the latter were large and welcome increases in IMF lending for LICs, two particular features of the IMF's response both during the immediate crisis and more recently have proved challenging. First, as pointed out above, the increases were far smaller for LICs than for MICs, and, more importantly, these facilities did not sufficiently compensate for the large scale of the external shocks. Second, some aspects of the more recent changes in IMF compensatory financing facilities have proved disappointing and seem, in several aspects, even to imply steps backwards.

More broadly, especially since the crisis started in 2007, reforms to IMF concessional financing facilities have put increased emphasis on shocks support. Such a change of emphasis, which has resulted in a greater proportion of IMF lending to LICs going to shocks support, is to be welcomed, although it is still insufficient in proportion to the magnitude of shocks during the period. This followed two decades in which the IMF's financial support to LICs was channelled mainly through three-year high-conditionality financial arrangements, and shocks were addressed by augmenting financing only under these arrangements. This greater emphasis for

LICs lies in contrast with the trend for MICs, which, unfortunately—for trade shocks, as we discuss below—has broadly been going in the opposite direction, reducing for these countries the importance of IMF compensatory financing for shocks. Nevertheless, the IMF has successfully implemented a facility for capital account shocks for MICs, which is positive.

The shift for LICs signals recognition by the IMF that the size, frequency and economic cost to the poor of external shocks tends to be higher in LICs than in other economies, increasing risk and uncertainty for private agents and governments, and that shocks can set back gains in increasing investment and growth as well as reducing poverty. Given their heavy reliance on commodity exports, LICs, particularly in Sub-Saharan Africa, experience median terms of trade volatility nearly twice as high as those in the rest of the world (IMF 2011). The new approach also recognises that, with improved macroeconomic policies and institutions and growing global integration of LICs, the importance of external shocks in driving output volatility has increased, compared with idiosyncratic domestic shocks, linked mainly to incorrect policies (Raddatz 2007).

Recognising improved macroeconomic management and the growing importance of short-term shock-related financing needs, reforms were undertaken that created two short-term financing instruments in January 2010: the Rapid Credit Facility (RCF) for emergency support and the Standby Credit Facility (SCF) for short-term quite high conditionality support. The SCF provides short-term financial arrangements and is applicable to shocks as well as many other circumstances. These new facilities replaced previously existing facilities. The aim of these changes was to streamline and simplify existing facilities and adjust them better to LIC needs (IMF 2009, 2011, 2012).

A broad question to ask is how much is gained through fairly small frequent changes in facilities that require a lot of effort in design by the IMF and understanding of changes by busy policy-makers in borrowing countries. It would seem far more worthwhile to make a significant change in terms of scale (so the lending would cover a bigger and more significant proportion of shocks), and a reduction or elimination of conditionality, which is not appropriate for external shocks or natural hazards; as we point out below, elimination of conditionality would also increase the speed of disbursement of IMF loans.

There were positive features in the reforms of the IMF LIC facilities. Consolidation simplified some of them, concessionality was increased and the RCF had longer maturity. IMF lending to small vulnerable economies was enhanced: often, such countries are also most at risk from disaster events. IMF emergency facilities were also consolidated in the RCF for post-conflict and natural disasters as well as external economic shocks, although the scale of lending, at only 25 % of the quota initially, was very small. Finally, the IMF streamlined conditionality for the SCF.

However, the new facilities have several shortfalls, especially for LICs. First, the original concept of IMF compensatory financing—as providing countries facing purely external shocks (whether they be exogenous economic or originating in nature) with almost automatic, very rapid liquidity constituting a significant proportion of the shock—continues to be sharply diluted. The only low-conditionality IMF shocks financing facility for LICs that remains is the RCF. At a level of 50 % of quota annually, with a total cumulative limit of 125 % in the case of external shocks, it is small (Berensmann and Wolf 2014). Only nine countries applied to this RCF new instrument during 2010–2013, for funds totalling around $230 million, implying rather modest scale in the use of this no-conditionality but small-scale instrument.

Furthermore, the CFF for terms of trade shocks for MICs, so widely and successfully used in previous decades (see detailed data in Griffith-Jones and Ocampo 2008), had previously been abolished. This is very problematic when terms of trade turn strongly against commodity exporters.

The IMF had explored for some time the creation of a preventive facility to deal with capital account shocks in MICs, which is valuable given increased importance of reversals of capital flows in those countries. Several attempts at creating instruments were not successful. In 2009, the IMF created the Flexible Credit Line for MICs, which it perceives as having very strong fundamentals but as risking facing capital account shocks. This facility was successful in that it has been used several times.

However, as regards to shocks on the trade account for MICs, there was a step backward from the purpose for which the IMF CFF was created (Goreux 1980: p. 3 [emphasis added]): "The facility would enable a member to borrow when its export earnings and financial reserves are low and repay when high, so its import capacity is unaffected by fluctuations in export earnings caused *by external events*". This was clearly based on the approach that IMF official liquidity should help avoid unnecessary negative effects on growth and poverty reduction. The CFF was created in 1963 as a low-conditionality facility to deal with external shocks relating to trade; through the years, there was a gradual increase in conditionality. From 2000, when CFF conditionality was raised to upper-credit tranche level, MICs stopped using it.

With a small-scale exception—the RCF—all compensatory financing for LICs took on upper-credit tranche conditionality. This both is inappropriate for external shocks and delays lending, making it thus less effective for its counter-cyclical role.

Limiting, or practically eliminating, low-conditionality shocks financing at the IMF both for LICs and MICs seems particularly undesirable in a world where external shocks are far more common as a result of frequent and increasingly global financial crises. Such an evolution of the world economy would seem to require more and especially tailored shocks financing, rather than far less and more diluted resources, as seems to emerge from some of the evolution of IMF facilities. Furthermore, more emphasis needs to be added to financing for disaster management.

Furthermore, IMF reforms of shocks financing are in contrast with overall positive trends, of lightening of structural conditionality at the IMF, reflected in the fact that, in 2009, the link between disbursements of IMF loans and performance on structural conditions was eliminated; and the somewhat greater emphasis on more counter-cyclical macroeconomic policies in light of the crisis. Greater commitments of IMF lending in general during periods of shocks is also very useful, even if it is not channelled through shocks facilities.

As a result of the crisis, there is growing consensus on the desirability of enhancing the predictability of shock financing, for instance by broadening options for contingent support, including making access to IMF resources automatic under certain circumstances. For example, for countries that have three-year IMF programmes, these could have an option for the country to request an automatic increase of the loan if certain economic external shocks—for example a reduction in their terms of trade by over 5 % or a certain natural disaster—take place. Even the scale of additional resources could be broadly stipulated *ex-ante*, linked to the potential magnitude of shocks. The IMF has in the past used such contingent clauses in very specific programmes; such a practice could be very beneficially expanded to, for example, all IMF three-year programmes. Ideally, the additional access would be less constrained by access limits linked to quotas and more closely linked to country needs.

Significantly increasing access to low-conditionality shocks, IMF facilities would also be desirable. This could most easily be done by significantly expanding the low-conditionality RCF for LICs.

To conclude on IMF financing, the response to the global financial crisis was important in terms of scale, and relatively speedy; it was better for MICs than LICs. However, even in MICs, the increase in IMF lending was far smaller than the initial contraction of private flows. In future, the shocks facilities need to expand—in terms of both scale and the more explicit inclusion of disasters originating from nature—and to become less conditional. The latter will also guarantee a quicker response, which will reduce negative impacts of shocks on investment, employment and poverty reduction. These shocks facilities need to be closely coordinated with those of other international institutions, such as the World Bank, to which we now turn.

7.4.2 The Response of the World Bank

Multilateral development banks (MDBs) rapidly increased lending commitments in response to the financial crisis. The World Bank almost doubled lending commitments, from $25 billion in 2008 to $47 billion in 2009 and $59 billion in 2010. Some RDB responses were also very large, with the African Development Bank increasing loan commitments by 137 % between 2008 and 2009 (see Griffith-Jones and Gottschalk 2012).

The World Bank delivered its response through four different mechanisms:

First, a pilot IDA Crisis Response Window (CRW) was created. For IDA 16, a permanent CRW was established with resources capped at 5 % of the total IDA 16 replenishment resources. This new permanent facility represents a more systematic approach for IDA in dealing with economic shocks and large natural hazards. The triggers for disbursement from the facility included:

(a) A projected decline of GDP growth of at least 3 percentage points in a significant number of IDA countries is required. This is a very stringent requirement as it has been empirically verified that few country projections of GDP growth reach a fall of 3 % or more; the probability of several LICs projecting such a sharp fall is even smaller.
(b) A key CRW objective is to protect core fiscal spending in the short term to avoid derailing long-term development objectives, which is very valuable. Therefore, fiscal indicators are included here.

The volume of the CRW was quite small, at just over $1.3 billion, or just over 4 % of IDA's envelope during the IDA 16 period. One important reason, as discussed above, for such a restricted allocation seems to be that the CRW—as currently conceived—binds scarce concessional resources that can be used for other purposes; however, this trade-off could be minimised if such resources could be used to build resilience against future likely shocks, particularly relevant for disasters originating in natural shocks.

Second, the approved the Immediate Response Mechanism (IRM), enabling LICs quick but limited access to funding after shocks, with emergency finance provided within weeks. IRM is not additional to agreed IDA country allocations, but is drawn from funds committed but not disbursed. It allows IDA countries to rapidly access up to 5 % of their undisbursed IDA investment project balances following natural disasters and severe economic shocks. For IRM to be effective, it is necessary for countries to incorporate contingent emergency parts in existing IDA investment projects. This, like augmentation arrangements in IMF programmes, seems positive and could be expanded (World Bank 2011).

Third, the World Bank Group set up the Global Food Crisis Response Programme to provide immediate relief to countries hard hit by high food prices. Between 2008 and 2010, $2 billion of World Bank funds was made available.

And fourth, the World Bank created the Rapid Social Response Programme to support LICs in social protection and access to basic social services.

Overall, MDBs responded substantially to the financial crisis. The crisis demonstrated the crucial counter-cyclical role they can play when shocks occur. While the international community had previously emphasised the role MDBs play in poverty reduction and provision of global public goods, this counter-cyclical role was not clearly recognised before. This meant many lessons from past experience were missed, which indicated that, aside from provision of liquidity during crises, it is equally important to provide official long-term finance when private finance dries up, or after natural hazards, and also to maintain the dynamics of investment. In addition, a very positive feature of the MDB response was that a number of targeted large regional initiatives were launched. The massive needs the crisis caused pushed these institutions to collaborate; such a fruitful approach could be applied more intensely to increasing resilience to natural hazards and financing disaster relief.

At the same time, however, a number of important factors constrained the scale and timeliness of the MDBs' response. In part, these constraints stemmed from limitations in the MDBs' capital. An important lesson is that there should be sufficient headroom available in the capital, and lending capacity, of MDBs, so they can respond quickly to shocks. As noted above, the response to the needs of LICs was also insufficient. Finally, the dynamics of rapidly expanding commitments were not reflected in disbursements, which for the World Bank grew far slower in 2008–2009 than the level of commitments.

In considering the role of MDBs in responding to future crises, there are several opportunities to strengthen responsiveness, some building on pilot and other limited initiatives trialled to date. There is strong scope for MDBs to introduce lending instruments that make developing countries less vulnerable during crises, either because they reduce currency mismatches by lending in local currency (which MDBs pioneered in some cases) or because they adjust the maturity of repayments of loans in a counter-cyclical manner, so net lending can increase more in bad times. An interesting mechanism would build on the successful example of Counter-Cyclical Loans used by the Agence Française de Développement, which provides debt holidays on its concessional loans to LICs that experience export shocks. This could be broadened to include shocks originating in nature.

7.5 Prevention, Risk Management and Co-benefits: Lessons for DRM

As a result of crises, there is growing consensus on the desirability of enhancing the predictability of shock financing, for instance by broadening options for contingent support, including making access to IMF resources automatic under certain circumstances. For example, for countries that have three-year IMF programmes, these could have an option for the country to request an automatic increase of the loan if certain external economic or natural shocks take place—for example a reduction of their terms of trade, by over 5 % or the occurrence of a certain natural disaster. Even the scale of additional resources could be broadly stipulated *ex-ante*, linked to the potential magnitude of shocks. The IMF has in the past used such contingent clauses in very specific programmes, as discussed above; such a practice could be very beneficially expanded to, for example, all three-year IMF programmes. Ideally, the additional access would be less constrained by access limits linked to quotas and more closely linked to country needs. By reducing uncertainty, such measures could lead to higher private investment, employment and growth.

Significantly increasing access to low-conditionality shocks facilities would also be desirable. This could most easily be done by significantly expanding the low-conditionality RCF for LICs. There would be a far better fit between the nature of the shock (external) and the instrument the IMF would use: a low-conditionality facility. As pointed out, a second key advantage would be greater speed in commitment of resources, which would increase the counter-cyclical nature of the lending instrument and avoid unnecessary costs to growth and poverty reduction. The smoothing of economic activity would also encourage higher and more sustained private investment, valuable for higher growth. More generally, the amount of financing provided by the World Bank, the RDBs and even the IMF was smaller than the initial contraction of private capital. Therefore, preventive measures, to avoid financial crises, need to be taken.

The examination of economic shocks presents clear parallels with the growing burden of disaster events and losses linked to changing hazard burdens and changing human exposure and vulnerability. A similar need therefore arises as in the case of economic shocks, to increase efforts and funding for DRM. From the development point of view it is essential to consider how a country can cope with different shocks occurring in close succession or simultaneously.

Te Velde et al. (2011) examine the impact of shock absorber schemes for Benin, Burundi, the Democratic Republic of Congo and Mauritius, especially since 2006. An important conclusion from the four case studies is that government spending, and government investment, as a proportion of GDP increased when shocks facilities were higher in response to shocks than what they would otherwise have been. These case studies show clearly the value for both short-term growth and long-term development of shock financing. There were some exceptions to this positive evolution, especially in countries particularly badly hit by external shocks or especially vulnerable to them. Countries that were hit by both external economic shocks and natural disasters seemed to suffer particularly strong declines of output and employment, as well as greater difficulties in recovering. This shows the significance of having strong and large international responses for both exogenous economic shocks and natural hazards.

A key policy lesson here is to ensure that both appropriate lending facilities and sufficient resources are in place before crises and other major shocks hit, and that shocks facilities can be disbursed quickly, requiring low conditionality and forward-looking triggers. Furthermore, it may be more appropriate to use special shocks facilities to provide most of the financing, rather than relying also a great deal on broad lending or grant mechanisms, as occurred in 2006–2011.

There are two further important features to highlight. First, the IFIs' broad response to the crisis was driven by a significant increase in overall lending, and much of the response was channelled through regular, rather than crisis, facilities. Notwithstanding this, shock financing through special facilities by IFIs also increased significantly for LICs, from very low levels in 2006–2008 to just over \$2.5 billion in 2010, the peak year. There seems to be a case for having in place *ex-ante* larger shocks facilities to respond rapidly and at a sufficient scale both to large exogenous economic shocks and to large natural disasters, as well as sufficient resources for such large shocks facilities to be rapidly disbursed. This would require greater commitment to grants for this purpose, to make higher concessional resources available for lending to LICs. In the case of the World Bank and the RDBs, this may require further increases of their capital as well, to allow them to borrow greater amounts on the capital markets once exogenous crises and/or natural disasters hit.

Similarly, the European Union (EU) could increase the proportion of its resources devoted to financing shocks facilities; these are currently made as grants to the ACP countries only. Two proposals are relevant here. One is to expand the countries covered to all developing countries, and not just former colonies of EU member countries, which practically all the ACP countries are; this would need to be accompanied by greater contributions from non-EU developed countries. Another proposal is to evaluate the possibility that funding provided to relatively less poor countries, especially if they have relatively low levels of debt, could be made through highly concessional loans, which would have the virtue of greater funding availability, covering a higher proportion of the shock (see Te Velde et al. 2011).

There are also compound links between disasters and financial crises. First, when natural disasters interact with external economic shocks, they seem far more damaging to short- and long-term economic prospects. As a consequence, they seem to require far higher international support. Second, natural disasters can be either frequent or one-off phenomena; furthermore, their effects can be either temporary or more permanent. As in the case of exogenous economic shocks, the latter distinction seems very important: if effects are temporary, international official liquidity may be the more appropriate instrument, whereas if the impact is more permanent, international official lending and especially grants may be more appropriate, particularly for poorer and more vulnerable economies.

An important policy question for both economic and disaster risk management concerns trade-offs in resource allocation. If more emphasis is placed on dealing with shocks, via specialised facilities, to increase the speed and scale of the response, how can the potential trade-off in allocating fewer funds to other development activities be addressed, as well as minimised? One way to reduce such a trade-off is to use the resources to increase resilience, especially if funds allocated for shocks for a certain period have not been used during most of the period. This is very relevant for IDA lending, as well as for EU resources, where funds per period and per country are broadly allocated. One example, in the field of trade—and the exposure to trade shocks by countries relying mainly on one or two export commodities—is to help fund investment in the diversification of the economy, especially in tradeables, so as to reduce the impact of falls in prices of specific commodities exported or rises in specific imports, such as food and energy. Such a policy of investment in diversification not only reduces the economy's vulnerability to shocks but also, by increasing resilience and opening new economic sectors to private and public investment, is likely to have additional development benefits, available even if shocks do not occur.

Furthermore, such allocation from international funding could be increased if the country itself is dedicating resources (e.g. through the public budget or as part of the plans of the ministry of the economy or of planning that encourage private investment towards such diversification). Such an approach is particularly relevant for disasters, given the special importance of increasing resilience *ex-ante*. Again, funds allocated to DRM for a certain period could, if no such disasters occur during most of that period, be allocated to investment in increased resilience to such disasters in the future. Examples could include investment in housing far from areas that are likely to be flooded.

Also, resources from abroad, such as from the World Bank, could be higher if the country itself were dedicating resources to building such resilience, for example through the budget (via the finance ministry) or via plans prepared by the ministry of planning or the environment to encourage private investment. Thus disbursement from such international lending facilities, when done *ex-ante*, could not just finance DRM through the resources lent or granted but also encourage finance and other ministries to devote more national resources to the important task of building resilience *ex-ante*. However, once a disaster happens, there should be no preconditions or conditionality for international disbursements, as time is of the essence to help rebuilding where appropriate, and more broadly to minimise damage to growth and poverty reduction.

Meanwhile, targeted large regional initiatives were launched as a response to the global financial crisis, mainly through joint collaborations among institutions, notably the World Bank working together with RDBs, but also with close coordination with the IMF. Examples are the Joint Plans in Africa, Latin America and the Caribbean and for Central and Eastern Europe. The massive needs the financial crisis caused pushed these institutions to collaborate rather than compete. A similar approach of close coordination and, where necessary, collaboration should be used in the future, and not only after major crises occur but also in programmes for building resilience *ex-ante*; this is especially relevant in investment to prevent natural hazards. Often, in such cases, it may be desirable to have collaboration both between international institutions and between them and regional programmes, as the most effective way of increasing resilience.

More broadly, it seems important to think in terms of a design of an integrated shocks architecture, which deals in a consistent fashion with both exogenous economic shocks and natural hazards, across the international and regional institutions providing loans or grants in the face of shocks. Alongside this, consideration should go to *ex-ante* enhancement of resilience that will both reduce vulnerability to shocks and promote long-term development. Such an integrated shocks architecture should be adequately funded, and should be permanent, so as to be able to disburse quickly when shocks or disasters hit. A review of existing facilities with a view to establishing a more permanent, well-coordinated, simple and financially sustainable shocks architecture has become important.

7.6 The Triple Co-benefits of DRM and Financial Risk Management

The case for the prevention of financial crises suggests greater national and international resources being channelled to disaster prevention and management would have not just immediate short-term benefits for growth and poverty reduction but also long-term benefits in terms of sustaining investment (both public and private) that is both disrupted by disaster events and inhibited simply by the risks of disaster. Financially, this economic investment is vital to helping sustain long-term development as well as long-term poverty reduction—a benefit that accrues regardless of the occurrence of a disaster.

The multiple benefits of financial risk management can therefore be seen in relation to the concept of the "triple resilience dividend" employed in DRM (Tanner et al. 2015). This argues that disasters do not cause devastation only in their aftermath; the risk of a disaster also can cause economic inefficiency and losses even before disaster strikes. While the benefits of avoiding losses and damages have been widely studied and documented, there has been far less focus on how investments in DRM can yield a real dividend, even in the absence of a disaster. Figure 7.1 highlights the triple dividend for investments in DRM. Reducing losses and damages in the event of a disaster is often the key motivating factor for DRM (first dividend).

However, even if the anticipated disaster does not occur for a long time, increased resilience means background risk is reduced and economic development potential is unlocked (second dividend). In addition to these primary objectives of DRM, investments in resilience may yield further social, economic and environmental co-benefits (third dividend). In the medium to long run, these benefits can trigger a wide range of benefits across society, income groups, geographic regions, government entities, industries and supply chains.

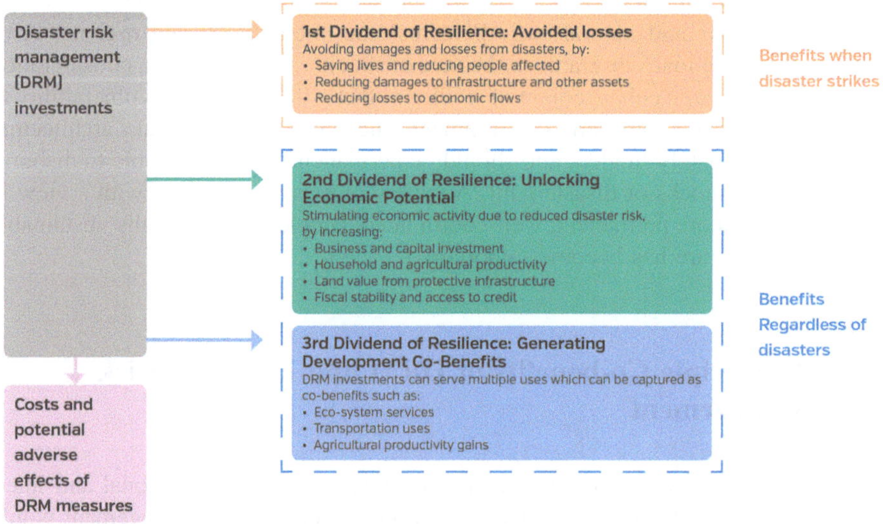

Fig. 7.1 The Triple Dividend of Resilience. *Source* Tanner et al. (2015); see chapter 1

7.6.1 First Dividend—Avoided Losses

For crisis management, this dividend relates to maintaining and enhancing growth, investment or employment in the face of economic shocks. Stability of growth is likely to lead to facilitate greater sustainability of fiscal policy and debt management.

Key factors in successful crisis management are centred on enhanced automaticity, which is reflected in both greater speed of commitments and disbursements and the large scale of purely automatic facilities. For the IMF, speed of disbursement relates to the ability to fund imports and prevent Balance of Payments crises by providing short-term official liquidity. For World Bank, MDB or EU disbursements, it is related to longer-term finance linked to maintaining investment in projects and sectors in the face of shocks, preventing business and investment interruptions that can be damaging to long-term growth.

7.6.2 Second Dividend—Reduced Background Risk

For economic resilience, an important way of reducing background risk is through regulatory measures that reduce the risk of future crises. Developing and emerging economies had generally become more cautious of the risks of financial crisis as a result of their past experiences, and more willing to introduce and implement financial regulation. Developed economies have been more willing to do so following the major crisis that started in 2007. Financial regulation can include increased capital, liquidity and leverage requirements, especially in the banking sector. It also includes separating within banks any activities relating to "regular" commercial activity from more risky speculative activities [e.g. following Vickers rule (in the UK) and Volker's rule (in the US)].

Reduced speculation makes it possible to use bank deposits increasingly to finance working capital and longer-term investment. This leads to higher, more efficient and more stable growth. It also reduces excessive risk-taking activity, leading to less pro-cyclical economic conditions, which will encourage private investment. However, financial regulation for stability has to be carefully designed to avoid negative impacts on longer-term investment and growth. There are concerns, for example, that tight regulation to demand higher liquidity for the insurance industry (done for prudential regulation purposes) may discourage the channelling of such funds into productive investments.

7.6.3 Third Dividend—Co-benefit of Financial Risk Management

Following from repeated financial crises, there has been growing recognition that a more diversified financial system can also reduce risks to stability and growth. In particular, there is growing acknowledgement of the valuable role national public development banks can play in providing counter-cyclical funding.

Furthermore, the funding of these banks on the private capital markets can help channel longer-term finance for funding national priorities. This can include the financing of public goods that otherwise may not take place, such as investments in DRM, climate change adaptation or renewable energy. Where such public goods have significant social or environmental externalities, loans from these banks can more easily be blended with public subsidies. In situations where fiscal resources are scarce, development banks also provide a good source of leverage of the public resources invested in their capital by raising finance in private capital markets as well as co-financing with private bank lending and private investment.

More broadly, prudent fiscal management will also assist in crisis prevention, as large fiscal deficits can be an important cause of financial crisis, as the case of Greece illustrates. Furthermore, the absence of financial crises helps maintain fiscal health, as crises are often extremely damaging to public revenues and add significant demands to public spending, for example bailing out banks and increased payments for benefits as unemployment increases (see Chap. 4).

7.7 Conclusions and Suggestions for Future Research

There are interesting parallels and lessons for DRM from financial crisis prevention and management. A particularly relevant one is the value of crisis prevention, as well as enhancing resilience *ex-ante* in the case that crises do occur. Both in the case of economic shocks, and financial crises in particular, and in DRM, valuable benefits from a development perspective can be accrued from such measures, not only if crises occur but also even if they do not, as the greater certainty will encourage higher investment as well as new economic opportunities. Naturally, it is crucial that *ex-ante* measures need to be complemented by sufficient and sufficiently speedy external economic and natural shocks *ex-post* compensatory facilities.

Further research is needed on the broad issues of potential trade-offs of devoting more resources *ex-ante* versus *ex-post*, but above all on how best to make both mutually complementary, while maximising both their effectiveness in avoiding the costs of natural or economic shocks to development and the positive impacts of resources deployed, under all circumstances. This broader understanding needs to be applied to the design of a shocks architecture, as well as effective mechanisms within it, in order to—in the most cost-effective way—maximise the impact on development, especially for poorer and more vulnerable countries. Flexibility built into regular mechanisms, speed of disbursement once shocks hit, accompanied by very low and appropriate conditionality, and the possibility of transferring resources from prevention to resilience are key criteria that need to be applied.

References

Aiello F (2009) Experiences with traditional compensatory finance schemes and lessons from flex. Working paper 12, Dipartimento di Economia and Statistica, Università della Calabria, Calabria

Atkinson T, Luttrell D, Rosenblaum H (2013) How bad was it? the costs and consequences of the 2007–8 financial crisis? Staff paper. Federal Reserve Bank of Dallas, Dallas, TX

Benson C, Clay E (2003) Economic and financial impacts of natural disasters: an assessment of their effects and options for mitigation. World Bank, Washington, DC

Berensmann K, Wolf P (2014) The role of international financial institutions in crises, improving the financial architecture of the World Bank and IMF in managing shocks. Discussion paper 33/2014, DIE, Bonn

Cornia G, Jolly R, Stewart F (1987) Adjustment with a human face. Oxford University Press, Oxford

Didier T, Hevia C, Sergio SL (2012) How resilient and countercyclical were emerging economies during the global financial crisis? J Int Money Finance 31(8):2052–2077

Dixit A, Pindyck R (1994) Investment under uncertainty. Princeton University Press, Princeton, NJ

Eichengreen B (2004) Global imbalances and the lessons of Bretton Woods. Working paper 10497, NBER, Cambridge, MA

Goreux LM (1980) Compensatory financing facility. Pamphlet series 34, IMF, Washington, DC

Griffith-Jones S (1983) Compensatory financing facility: a review of its operation. G24, Washington, DC

Griffith-Jones S, Gottschalk R (2007) Asian tigers, new vulnerabilities to crisis? IDS Bull 38(4)

Griffith-Jones S, Gottschalk R (2012) Exogenous shocks—dealing with the only certainty—uncertainty. Paper for commonwealth finance ministers meeting

Griffith-Jones S, Ocampo JA (2008) Compensatory financing for shocks: what changes are needed? Paper prepared for UN Committee on development policy

Griffith-Jones S, Ocampo JA (2009) The financial crisis and its impact on developing countries. Working paper 53, IPC-IG, Brasilia

Harper C, Jones N, McKay A, Espey J (2009) Children in times of economic crisis: past lessons, future policies. Background note. London, ODI

IMF (International Monetary Fund) (2009) A new architecture for facilities for low income countries. IMF, Washington, DC

IMF (International Monetary Fund) (2011) Managing volatility in low-income countries: the role and potential for contingent financial instruments. IMF, Washington, DC

IMF (International Monetary Fund) (2012) Review of facilities for low income countries. IMF, Washington, DC

Olsson L, Opondo M, Tschakert P, Agrawal A, Eriksen SH, Ma S, Perch LN, Zakieldeen SA (2014) Livelihoods and poverty. In: Field C, Barros V, Dokken DJ, Mach KJ, Mastrandrea MD, Bilir E, Chatterjee M, Ebi K, Estrada Y, Genova RC, Girma B, Kissel ES, Levy AN, MacCracken S, Mastrandrea PR, White LL (eds) Climate change 2014: impacts, adaptation, and vulnerability, Part A: global and sectoral aspects, Cambridge University Press, Cambridge and New York

Raddatz C (2007) Are external shocks responsible for the instability of output in low income countries? J Dev Econ 84(1):155–187

Reinhart C, Rogoff K (2008) The aftermath of financial crises. American economic association meeting, San Francisco, CA, 9 Jan

Rodin J (2014) The resilience dividend: being strong in a world where things go wrong. Public Affairs, New York

Tanner T, Surminski S, Wilkinson E, Reid R, Rentschler J, Rajput S (2015) The triple dividend of resilience: realising development goals through the multiple benefits of disaster risk management. Global Facility for Disaster Reduction and Recovery (GFDRR) at the World Bank and Overseas Development Institute (ODI), London

Te V, Willem D, Griffith-Jones S, Kingombe C, Kennan J, Tyson J (2011) Study on shock absorbing schemes in ACP countries. Final report for the EC

UNISDR (UN Office for Disaster Risk Reduction) (2015) Making development sustainable: the future of disaster risk management. Global assessment report on disaster risk reduction, UNISDR, Geneva

Vorhies F (2012) The economics of investing in disaster risk reduction. Working paper, UNISDR, Geneva

Wilkinson E (2012) Transforming disaster risk management: a political economy approach. ODI, London

Winters LA, Lim W, Hanmer L, Augustin S (2010) Economic growth in low income countries: how the G20 can help to raise and sustain it. ED-WPS 8-2010, IDS, Brighton

World Bank (2011) The world bank group's response to the global economic crisis, phase II. IEG, Washington, DC

World Bank (2013) World development report 2014. Risk and opportunity, Managing risk for development, World Bank, Washington, DC

Index

S. Surminski and T. Tanner (eds.), *Realising the 'Triple Dividend of Resilience'*, Climate Risk Management, Policy and Governance,
DOI 10.1007/978-3-319-40694-7

MIX
Papier aus verantwortungsvollen Quellen
Paper from responsible sources
FSC® C105338

If you have any concerns about our products,
you can contact us on
ProductSafety@springernature.com

In case Publisher is established outside the EU,
the EU authorized representative is:
Springer Nature Customer Service Center GmbH
Europaplatz 3, 69115 Heidelberg, Germany

Printed by Libri Plureos GmbH
in Hamburg, Germany